Sequence alignment and phylogenetic inference

Inaugural-Dissertation

zur

Erlangung des Doktorgrades der

Mathematisch-Naturwissenschaftlichen Fakultät

der Heinrich-Heine-Universität Düsseldorf

vorgelegt von

Roland Fleißner

aus München

Logos Verlag Berlin

2004

Bibliografische Information Der Deutschen Bibliothek

Die Deutsche Bibliothek verzeichnet diese Publikation in der Deutschen
Nationalbibliografie; detaillierte bibliografische Daten sind im Internet über
http://dnb.ddb.de abrufbar.

ISBN 3-8325-0588-1

Logos Verlag Berlin
Comeniushof, Gubener Str. 47,
10243 Berlin
Tel.: +49 030 42 85 10 90
Fax: +49 030 42 85 10 92
INTERNET: http://www.logos-verlag.de

Gedruckt mit der Genehmigung der Mathematisch-Naturwissenschaftlichen

Fakultät der Heinrich-Heine-Universität Düsseldorf

Referent: Prof. Dr. Arndt von Haeseler

Koreferenten: Prof. Dr. Gerhard Steger, Prof. Dr. Rolf Backofen (Jena)

Tag der mündlichen Prüfung: 18. Dezember 2003

Premature optimization is the root of all evil.

(Donald Knuth)

Acknowledgements

I would like to thank my advisor Arndt von Haeseler for having the idea for this project and for his patience, Dirk Metzler and Anton Wakolbinger for their enthusiasm and their expertise, Gunter Weiß for explaining basic statistics, Bojan Basrak and Thomas Schlegel for listening to my stories, Heiko Schmidt and Achim Radtke for giving advice on the usage of computers, Lutz Voigt for keeping the system running and Tanja Gesell for stimulating discussions. I also thank everybody at the zoological institute in Munich, the MPI for evolutionary anthropology in Leipzig, the mathematics department in Frankfurt and at the bioinformatics institute in Düsseldorf for the pleasant working environments. I am indebted to the Deutsche Forschungsgemeinschaft and the Max Planck Gesellschaft for financial support and to my friends and especially my family for moral back up.

Publications

Parts of this study have already been published or have been submitted for publication:

1. Roland Fleißner, Dirk Metzler and Arndt von Haeseler. 2000. Can one estimate distances from pairwise sequence alignments? In: *GCB 2000, Proceedings of the German Conference on Bioinformatics, October 5-7, 2000, Heidelberg* (Bornberg-Bauer, E., Rost, U., Stoye, J., and Vingron, M., eds) pp. 89–95, Logos Verlag Berlin.

2. Dirk Metzler, Roland Fleißner, Anton Wakolbinger and Arndt von Haeseler. 2001. Assessing variability by joint sampling of alignments and mutation rates. *J. Mol. Evol.* **53**:660–669.

Contents

Introduction

When studying a set of related nucleotide or amino acid sequences, we are interested in several questions: What does the underlying tree look like? Which positions in the sequences are homologous? What are the characteristics of the evolutionary process that led to these sequences? How reliable are our deductions? We normally try to tackle these questions by first constructing an optimal multiple alignment on which we then base all our inference on the tree topology and the parameters of the substitution model we prefer. Due to highly elaborate alignment (e. g. Thompson *et al.*, 1994) and tree reconstruction programs (e. g. Swofford, 1991; Felsenstein, 1993; Strimmer and von Haeseler, 1996), this way of proceeding is fast and certainly works well in most cases. Nevertheless, it has several drawbacks: Firstly, as we did not attempt to model the insertion and deletion process explicitly, we can hardly make any statements about the nature of this process. Secondly, in order to find an optimal alignment we have to specify mismatch and gap penalties, thus making implicit assumptions about the plausibility of substitutions, insertions and deletions. Thirdly, multiple alignment methods either align the sequences according to a prespecified phylogenetic tree or ignore their evolutionary history (cf. Gotoh, 1999). Either way, our estimate of the phylogeny will be affected. Finally, our knowledge of the statistics of alignment scores and of the behaviour of the heuristics which are used to come close to the

1

optimum is at the moment very limited. Thus, the problem of phylogenetic analysis is twofold: Our estimates of phylogenies and model parameters are influenced by the multiple alignments on which we base our studies, yet we do not know how to judge alignment errors and thus have no clue about the reliability of our estimates.

This thesis deals with some of these problems arising at the junction between sequence alignment and phylogeny reconstruction: After giving a brief overview over some basic concepts and methods of phylogenetic analysis in chapter 1, we will study the influence of global sequence alignment on the estimation of pairwise distances (chapter 2) as well as on the reconstruction of phylogentic trees (chapter 3). Chapter 4 gives an introduction into the application of insertion and deletion models. These models are then used to simultaneously reconstruct phylogenies and multiple alignments (chapter 5). Finally, we describe in the appendices the usage and some implementation details of the main programs that were written for this thesis as well as some possible generalizations of the models applied.

Chapter 1

Basics of phylogenetic inference

The problems dealt with in this thesis originate from the study of phylogenetic relationships on the basis of biological sequences. In this chapter, we
give a brief introduction into the field and its terminology. At the center of
all our studies are sets of related biological sequences. These sequences may
be either DNA sequences, proteins or all kinds of RNAs, but for the purpose
of this thesis they are simply thought of as strings whose letters belong to the
canonical alphabets of nucleotides or amino acids (cf. Voet and Voet, 1995).
Thus, we are ignoring any biochemical features of nucleotides or amino acids,
and when we speak of DNA, we will always mean one strand of the double
helix. We consider sequences to be 'related' or 'homologous' if they derive
from a common ancestor. Due to the way that DNA is replicated, all the
sequences which stem from one ancestral sequence are related by a bifurcating tree: One sequence replicates and produces two daughter sequences,
each of these daughter sequences replicates after some time and, in turn,
produces two daughter sequences and so on. Normally we do not have the
whole progeny of an ancestral sequence at hand but only a small subset. Yet,
since picking a subset of the progeny corresponds to pruning the tree, the

3

relationship of this subset's sequences is still adequately described by a bi-furcating tree which is called the sequences' phylogeny or phylogenetic tree. Genomic rearrangements like recombinations, inversions or transpositions do however obscure this tree-like structure or complicate sequence comparison by destroying the sequences' original order. Throughout this thesis we will therefore assume that these rearrangement events have already been sorted out and we will only focus on sequence changes that occur by substitutions[1], insertions or deletions. In the following, we describe some of the methods which lead from such data sets of unaligned biological sequences to estimates for the underlying trees.

1.1 Sequence alignment

In order to be able to apply the tree reconstruction methods that are introduced in the next section we have to know which bases of the sequences are homologous and which ones are only present in a subset of the sequences due to insertion or deletion events. We will depict any assertions about the homology or the lack of homology between bases of different sequences in the form of sequence alignments. A sequence alignment is simply an array where each row corresponds to one of the sequences and where those bases which are assumed to be homologous to each other stand in the same column (see figure 1.1). We get such an alignment by inserting special characters — throughout this thesis we will use the underscore — into the sequences so as to make all of them have the same length. In the case that our alignment covers the entire sequences we speak of a global alignment. If we are only making assumptions about the relatedness of some parts of the sequences

---

[1]In this thesis we will use 'substitution' as synonym for 'point mutation'.

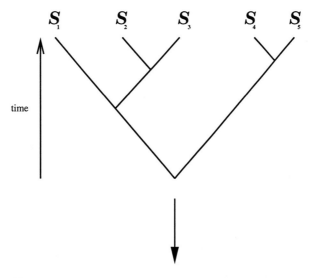

S_1 ATAACACAT_ATTACAAGCAAGTACCCCCCC
S_2 GTAACACAT_ATTACAAGCAAGTACCCCCCC
S_3 GCAGTACAT_A_____ACCCCCCC
S_4 GTAGTACAT_ATTACAAGCAAGCA_CAGCAA
S_5 ATAACACATAATTACAAGCAAGTA_CAGCAA

S_1 ATAACACATATTACAAGCAAGTACCCCCCC

S_2 GTAACACATATTACAAGCAAGTACCCCCCC

S_3 GCAGTACATAACCCCCCC

S_4 GTAGTACATATTACAAGCAAGCACAGCAA

S_5 ATAACACATAATTACAAGCAAGTACAGCAA

Figure 1.1: Top: A phylogenetic tree depicts the relationship of a set of sequences. Center: The homology relations of the individual bases that result from sequence evolution via point mutations, insertions and deletions can be represented as a multiple sequence alignment. Bottom: The observables of this process are just the unaligned sequences.

we call it local alignment. An alignment which only has two rows is called a pairwise alignment. If it has more than two rows we call it a multiple alignment. A whole discipline of bioinformatics is focusing on the development and the study of algorithms that find the most plausible alignment for a set of sequences. Before we describe how this is done for multiple alignments we treat the case of pairwise alignments.

A simple way to judge a global pairwise alignment A is given by the Needleman-Wunsch scoring function (Needleman and Wunsch, 1970):

$$S_{NW}(A) = \alpha a(A) + \beta b(A) + \gamma c(A). \qquad (1.1)$$

Here, $a(A)$ is the number of matches in the alignment A, i. e. the number of columns of A where the base in the first row coincides with the one in the second row, $b(A)$ is the number of mismatches, i. e. the number of columns where the bases in the two rows differ, and $c(A)$ is the number of gaps, i. e. the number of columns which contain a gap symbol ('_') either in the first or in the second row. The parameters α, β and γ are called the match, mismatch and gap score respectively and are used to weigh the plausibilities of match, mismatch and gap events. Usually β is chosen to be smaller than α, and γ is set to a value smaller than $\beta/2$, i. e. a match scores higher than a mismatch and a mismatch scores higher than two gaps. An optimal Needleman-Wunsch alignment is an alignment where $S_{NW}(A)$ reaches its maximum and which therefore maximizes the sequences' similarity. For two sequences x and y of length n and m respectively there are however already $\binom{m+n}{n}$ traces (cf. Ewens and Grant, 2001), i. e. groups of alignments which have the same combination of aligned residue pairs although different orderings of the indels. Since this number is growing very fast with n and m, it is not feasible to find the maximum of $S_{NW}(A)$ by computing the score of every thinkable alignment. Yet, it is also unnecessary. Suppose we drew a little arrow on the grid of integers $\mathbb{Z}^2$ for every position of a global alignment of x and y. Suppose also that each of these arrows started at the grid point whose coordinates are given by the lengths of the prefixes of x and y which contain every base to the left of the respective alignment position and that it ended in the grid point which corresponds to the lengths of x's and y's prefixes which end immediately after the bases of the respective alignment position. It is easy to see that what we would get is a directed acyclic graph with the following properties: it starts at grid point $(0,0)$, ends in grid point (n,m) and the coordinates of its nodes increase monotonically and in steps of at most 1 (see figure 1.2Ⓑ). We will call such an object the path representation of an alignment or simply an alignment path. The union of all possible alignment paths of two sequences yields the so-called alignment graph (see figure 1.2Ⓒ). The structure of this alignment graph now suggests finding the optimal Needleman-Wunsch alignment via the following dynamic programming[2] algorithm (Gotoh, 1982): Let $S_{max}(i,j)$ be the score of the optimal Needleman-Wunsch alignment of the first i positions of x and the first j positions of y. Obviously, $S_{max}(n,m)$ is the score of the optimal Needleman-Wunsch alignment of the entire sequences. Like every other alignment path of x and y the optimal one will end in node (n,m) of the alignment graph and it will have entered this node coming from one of this node's three neighbours $(n-1,m)$, $(n,m-1)$ and $(n-1,m-1)$. In the latter case the last edge of the alignment path corresponds to a match or a mismatch depending on the last bases of x and y whereas the first two cases stand for alignments which end with a gap in y or in x respectively. Thus, we get the following equation for the score of the optimal alignment

---

[2]Dynamic programming (Bellman, 1957) is an optimization strategy that decomposes large problems into sub-problems whose optimal solutions are already known.

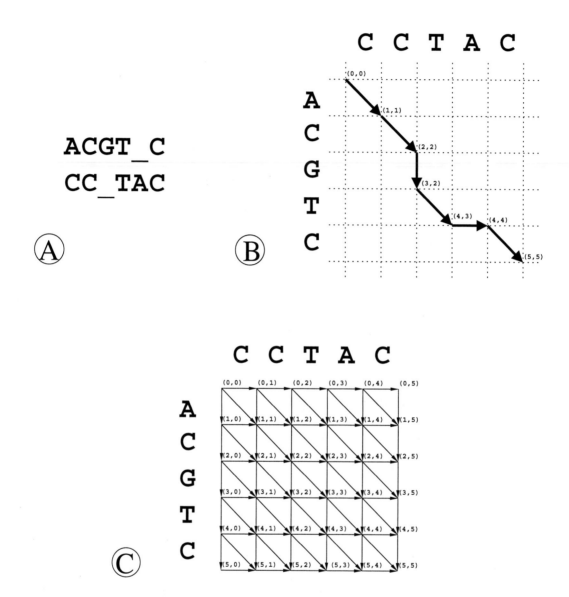

Figure 1.2: The graphical representation of pairwise alignments. Ⓐ: A global alignment of the two sequences ACGTC and CCTAC. Ⓑ: The corresponding alignment path. Ⓒ: The two sequences' alignment graph.

$$S_{max}(n,m) = \max \begin{cases} S_{max}(n-1, m-1) + s(x_n, y_m), \\ S_{max}(n-1, m) + \gamma, \\ S_{max}(n, m-1) + \gamma, \end{cases} \qquad (1.2)$$

where γ is the gap score and $s(x_i, y_j)$ is either equal to the match score α if the bases x_i and y_j match or to the mismatch score β if they differ. As the same reasoning is true for any $S_{max}(i, j)$ with $1 \leq i \leq n$ and $1 \leq j \leq m$ and as we can set $S_{max}(0,0) = 0$, $S_{max}(i,0) = i\gamma$ and $S_{max}(0,j) = j\gamma$, $S_{max}(n,m)$ can easily be computed by evaluating the $S_{max}(i,j)$ row by row and in each row from left to right. While computing $S_{max}(i,j)$ for some node (i,j) we also highlight the edge of the optimal path that enters this node. Thus we do not only get the optimal score $S_{max}(n,m)$, but we can trace back from (n,m) along the highlighted path and derive the associated optimal alignment.

Dynamic programming recursions like the one described above form the core of the search for optimal alignments. Similar algorithms exist for pairwise local alignments (Smith and Waterman, 1981), for scoring functions with a more elaborate treatment of gaps (Altschul and Erickson, 1986; Miller and Myers, 1988) and also the Viterbi algorithm, which is used to find the most probable state path through a hidden Markov model (see chapter 4), works like this. In principle the problem of aligning more than two sequences can be addressed in the same way. Every additional sequence will however add one dimension to the grid in which the alignment graph is embedded. Therefore, dynamic programming becomes impractical even for a small number of sequences.

Another problem with multiple alignment lies in the choice of an appropriate scoring scheme. The easiest way to assign a score to a multiple alignment M is the so-called sum-of-pairs objective function which simply yields the sum of the scores of all the pairwise alignments that are induced by M. When we compute the sum-of-pairs score of a multiple alignment we do however ignore the relationship of the sequences. Weighted sum-of-pairs scoring schemes (Altschul and Lipman, 1989) try to attenuate this problem by weighing the pairwise alignment scores according to the evolutionary distance of the respective sequence pair. Based on a sum-of-pairs scoring scheme, Carrillo and Lipman (1988) developed an algorithm which successfully excludes regions of the alignment graph that need not be considered in the search for the optimal multiple alignment. Thus, one can find the optimal sum-of-pairs alignment for up to seven protein sequences of 200 to 300 residues (Lipman *et al.*, 1989). If one wants to align more and longer sequences, as is usually the case in phylogenetic studies, one has to rely on heuristics.

One commonly used and very fast strategy is the so-called progressive alignment, which was put forth by a number of papers (e. g. Waterman and Perlwitz, 1984; Feng and Doolittle, 1987; Higgins and Sharp, 1989) and which also forms the basis of the widespread alignment program CLUSTALW (Thompson *et al.*, 1994). Progressive alignment algorithms construct a multiple alignment through several pairwise alignment steps. First, a pair of sequences is aligned and then one sequence after the other is added. Depending on the implementation, this adding of sequences is done either by standard pairwise alignments or by profile[3] alignments. The input order of the sequences during this procedure is however known to influence the alignments (Lake, 1991). Moreover, this input order is usually determined by a guide tree which is constructed from pairwise distance estimates and which might then bias the shape of the trees that are estimated based on these

---

[3]Profiles (Gribskov *et al.*, 1987) are sequences of frequency vectors which hold for every column of a multiple alignment the frequency that each character appears in that column.

multiple alignments. Progressive alignment algorithms will also stick to suboptimal decisions once they have been made. Therefore, a number of iterative refinement strategies have been developed (e. g. Barton and Sternberg, 1987; Berger and Munson, 1991). These strategies start from a preliminary multiple alignment. This alignment's sequences are then repeatedly divided into two groups whereupon the alignments within the groups stay fixed and the alignment between the two groups is computed anew.

For a more detailed introduction into the field see Durbin *et al.* (1998), Gotoh (1999), Gusfield (1997) or Waterman (1995).

1.2   Reconstructing phylogenetic trees

Algorithms abound that yield a phylogenetic tree given a multiple alignment of sequences. They differ in the way they price the possible trees as well as in the procedure with which they search for the optimal topology. For a detailed overview over most of the methods that have been proposed see Swofford *et al.* (1996). In this section we will describe only two of the optimality criteria that are in use, namely the minimum evolution criterion and the maximum likelihood principle.

Minimum evolution (Rzhetsky and Nei, 1992):

Suppose we know a way to compute a distance for every pair of aligned sequences x and y, i. e. there is a function d which gives a nonnegative real number $d(x, y)$ that fulfils the following criteria: $d(x, y) = 0$ if and only if $x = y$ (distinctness), $d(x, y) = d(y, x)$ (symmetry), and $d(x, y) \leq d(x, z) + d(z, y)$ for any third sequence z (triangle inequality). The problem of reconstructing a phylogeny given the pairwise distances for a set of n sequences is then

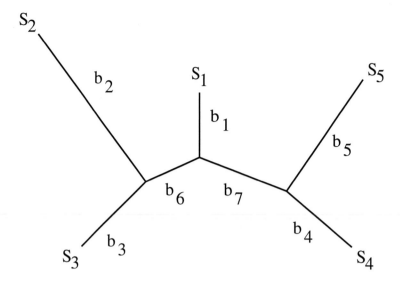

Figure 1.3: A phylogeny which shows the relatedness of five sequences (S_1 to S_5). The b_i symbolize the branch lengths. The distance between any two sequences can be computed by summing up the lengths of the branches which connect them. As this tree does not contain a node which corresponds to the ancestor of the five sequences, it is called an unrooted tree. See the top of figure 1.1 for an example of a rooted tree.

equivalent to arranging n points on a plane and finding a tree-shaped graph which has these points as leaf nodes[4] and where the length of a path that connects any two of these leaves is proportional to the distance between the corresponding sequences (see figure 1.3). This is only possible if the distance measure that we have chosen fulfils the so-called four-point inequality (Bunemann, 1971), i. e. for any four sequences a, b, c and d

$$d(a,b) + d(c,d) \ \leq \ \max(d(a,c) + d(b,d), d(a,d) + d(b,c)) \qquad (1.3)$$

where max is the maximum value function. Distance functions for which condition 1.3 holds are called additive distances. This additivity will, however,

---

[4]Nodes which have only one adjacent edge are called external nodes or leaves. Nodes with more adjacent edges are called internal nodes.

be violated by any real data set. Hence, instead of simply drawing a tree which represents the data's distance matrix we will have to fit a tree to our pairwise distances. This fitting is usually done with least squares methods (Cavalli-Sforza and Edwards, 1967; Fitch and Margoliash, 1967). Let T be some unrooted phylogeny relating n sequences and let $\hat{b}_k(T)$ be the estimated length of the k-th branch of T that we got by fitting T to the pairwise distances of the n sequences. The length $L(T)$ of T is then defined to be the sum over all branch lengths of T:

$$L(T) := \sum_{k=1}^{2n-3} \hat{b}_k(T) \qquad (1.4)$$

Rzhetsky and Nei (1992) suggested to take the tree for which $L(T)$ is minimal as estimate for the sequences' phylogeny. This tree is usually called the minimum evolution tree. If the chosen distance measure is additive and if we can obtain perfect estimates of the pairwise distances, the minimum evolution criterion will pick the correct tree. Finding the minimum evolution tree is however quite costly. For n sequences there are $\prod_{i=3}^{n}(2i-5)$ possible binary unrooted tree topologies (Felsenstein, 1978a). An exhaustive search for the minimum evolution tree would involve getting least squares estimates of the branch lengths for each one of them. Therefore, heuristics like the neighbor-joining method (Saitou and Nei, 1987, see also appendix B) which yield branch length estimates and a tree topology simultaneously are normally applied when one reconstructs a phylogeny from pairwise distances.

Maximum likelihood:

Statistics knows several methods to estimate model parameters from given data. One of them is the extensively used maximum likelihood estimation (cf. Edwards, 1972). Let x symbolize the data which is observed in some experiment and assume that x is a realisation of a random variable X whose distribution depends on some parameter θ. The basic task of statistical inference is to estimate θ from the data x. Maximum likelihood's answer to this problem is to take the θ as estimate which maximizes the probability to observe the data. This probability of x given θ is also called the likelihood of θ given the data.

Felsenstein (1981) adapted this method to the reconstruction of phylogenetic trees. In maximum likelihood tree reconstruction the phylogeny's topology and its branch lengths are seen as parameters of a stochastic process that generated the sequences. Provided that we model this process appropriately, maximum likelihood tree estimation is known to yield the correct tree if only we have enough data (Rogers, 1997). Yet, just like in the case of minimum evolution, an exhaustive search through tree space is infeasible even for a moderate number of sequences. There are however several heuristic search strategies which apply maximum likelihood estimation (Olsen *et al.*, 1994; Strimmer and von Haeseler, 1996).

Maximum likelihood tree reconstruction requires explicit modelling of sequence evolution. We will, for the moment, ignore the insertion-deletion process (see chapter 4) and focus only on the substitutional changes. To keep the computation tractable one commonly makes the assumption that the substitution process acts independently and in the same mannner on every branch of the tree. It is further assumed that each of the alignment positions evolves independently of the others, yet according to the same process. Thus, the evolution of sequences on a tree can be described in terms of the substitutional changes of single positions during some timespan t (Felsenstein, 1981). If one wants to model the substitution process which takes one nucleotide or amino acid i into another one (j) after time t, one has to be able to compute the probability of the initial state i as well as the conditional probability of j given i for every positive t. Although this might prove difficult in the general case, it gets easy when this substitution process is a stationary, time-homgeneous and reversible Markov process (cf. Tavaré, 1986). To describe such a process one only has to specify a rate matrix Q, where each entry Q_{ij} stands for the rate of change from state i to state j during an infinitesimal period of time. Since the process is time-homogeneous, one can then compute the transition probability matrix $P(t)$, whose entries $P_{ij}(t)$ represent the probabilities to be in state j after time t given that the initial state was i, as matrix exponential e^{Qt}. Due to stationarity, the frequencies π_i of the nucleotides or amino acids do not change over time and as the process is reversible one does not have to bother about the placement of the root. Table A.1 in the appendix lists the rate matrices of some common nucleotide substitution models.

The described substitution models are not only a prerequisite of maximum likelihood tree reconstruction but the processes' expected number of substitutions per positions constitutes a distance measure for which condition 1.3 holds and which may therefore serve as an input to the distance-based tree reconstruction methods. See Zharkikh (1994) on how to estimate this quantity from the observed sequence differences in a pairwise alignment.

Chapter 2

Distance estimation from pairwise sequence alignments

2.1   Introduction

In this chapter, we will have a closer look at global pairwise alignments. Think of two homologous sequences x and y and of the set $\mathcal{A}$ which is to contain all possible global alignments of these two. As the two sequences are homologous and as we are assuming evolution to proceed only via insertions, deletions or substitutions (see chapter 1) there is a nonempty set $\mathcal{A}_{true} \subset \mathcal{A}$ of alignments which are in accordance with the evolutionary history of x and y, i. e. alignments which align exactly those positions that are related by a series of substitutional events and where exactly those positions are aligned to a gap symbol that do not have a homologous counterpart in the other sequence. We will call any $A_{true} \in \mathcal{A}_{true}$ a true alignment of x and y. It is these true alignments that global alignment algorithms are trying to recover through maximizing the sequences' similarity. Figure 2.1 sketches the difficulty of this task: In that figure only the edges of the alignment graph which correspond

17

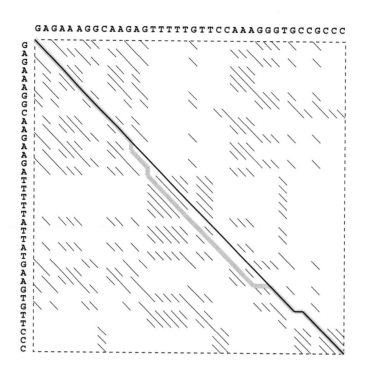

Figure 2.1: A toy example illustrating possible differences between the true alignment (grey path) and an optimal global alignment (black path). The thin lines mark possible matches and the dashed rectangle signifies the contour of the alignment graph.

to a match are drawn. Finding an optimal alignment means finding an alignment path which picks up as many of the marked edges as possible yet tries to avoid vertical, horizontal and unmarked diagonal edges. As there is no such optimality criterion for true alignments, an optimal alignment path may well deviate from the true ones. Furthermore, whether an alignment path is optimal or not depends on the parameters of the scoring function (see section 1.1). Thus, different parameter values may lead to different paths (see below).

In the last decade several articles have investigated the fidelity of optimal sequence alignments, i. e. the fraction of diagonal edges of a true alignment that are also covered by the optimal path[1]. Holmes and Durbin (1998) were able to show that given the probabilities of mismatches and gaps in the true alignments one can choose the parameters of the Needleman-Wunsch scoring function (see section 1.1) so as to maximize the fidelity of the optimal alignment. They also demonstrate that this maximal fidelity decreases the more the sequences are diverged. By applying statistical physics methods Hwa and Lässig (1996) claim that for long sequences the fraction of correctly found matches does not only tend to zero with increasing sequence divergence but that there is also a threshold above which none of a true alignment's match positions can be detected. Mevissen and Vingron (1996) developed a method to assign reliability values to individual alignment positions and come to the conclusion that for highly diverged sequences even those alignment positions with a high reliability value may be erroneous.

Although these are important results, alignment fidelity is not the quantity one is interested in when one wants to use optimal alignments in a phylogenetic context. For phylogenetic inference, the most important quantity to be estimated for a pair of sequences is the number of substitutions per position that they have undergone during the evolution from their common ancestor (see section 1.2). If we assume that the substitution process acts independently and identically distributed on each site and if we further assume that all nucleotides have the same frequency and that all substitutions occur at the same rate, then there is a simple way to estimate this number from the amount of differences per position in the true alignments of the two sequences (Jukes and Cantor, 1969). The number of differences per position

---

[1]Note that although there may be more than one true alignment the fidelity of a given optimal alignment is a well-defined quantity as two true alignments may only differ in the ordering of adjacent gaps.

in the true alignments, which we will also call the true Hamming distance per position, is defined as

$$d(\mathcal{A}_{true}) := \frac{b(\mathcal{A}_{true})}{a(\mathcal{A}_{true}) + b(\mathcal{A}_{true})}. \qquad (2.1)$$

In the above definition, $a(\mathcal{A}_{true})$ is the number of matches in a true alignment and $b(\mathcal{A}_{true})$ its number of mismatches. However, since $\mathcal{A}_{true}$ is unknown so is $d(\mathcal{A}_{true})$. Hence we cannot compute the number of substitutions per position from $d(\mathcal{A}_{true})$, but we may try to do so starting from

$$d(A_{opt,\theta}) := \frac{b(A_{opt,\theta})}{a(A_{opt,\theta}) + b(A_{opt,\theta})} \qquad (2.2)$$

which is the observed Hamming distance per position in an alignment $A_{opt,\theta}$ that is optimal for some tuple θ of alignment parameters.

For the sake of convenience, we will henceforth write d instead of $d(\mathcal{A}_{true})$ and $\hat{d}_\theta$ instead of $d(A_{opt,\theta})$. We will also refer to the numbers of matches, mismatches and gaps in the true alignments as a, b and c instead of $a(\mathcal{A}_{true})$, $b(\mathcal{A}_{true})$ and $c(\mathcal{A}_{true})$ and we will denote the respective numbers in $A_{opt,\theta}$ as $\hat{a}_\theta$, $\hat{b}_\theta$ and $\hat{c}_\theta$.

2.2 The observed distance of optimally aligned sequences

As described above, the true alignments and the optimal alignment of a sequence pair may differ and so may d and $\hat{d}_\theta$. Furthermore, as the optimal alignment depends on θ so does $\hat{d}_\theta$. In this section, we want to answer the question which values of $\hat{d}_\theta$ can be observed given the numbers of matches, mismatches and gaps of the true alignments and given θ. Since true alignments for real sequences are in general unknown we addressed this problem

20

through simulation. We restricted ourselves to the case of sequences of the same length with equal base frequencies and with equal frequencies of the 12 types of mismatches that are possible for pairs of DNA sequences.

Case	Scoring function	$\lambda \in$	Behaviour
I: $\alpha > \beta$	maximize $S_\lambda(A) = -b(A) - \lambda c(A)$ with $\lambda = \frac{\alpha/2 - \gamma}{\alpha - \beta}$	$(-\infty, 0)$ $[0, \frac{1}{2})$ $[\frac{1}{2}, \infty)$	only gaps no mismatches less and less gaps for $\lambda \to \infty$
II: $\alpha < \beta$	minimize $S_\lambda(A) = -b(A) - \lambda c(A)$ with $\lambda = \frac{\alpha/2 - \gamma}{\alpha - \beta}$	$(-\infty, 0]$ $(0, \frac{1}{2}]$ $(\frac{1}{2}, \infty)$	less and less gaps for $\lambda \to -\infty$ no matches only gaps
III: $\alpha = \beta$	maximize $S_\lambda(A) = -\lambda c(A)$ with $\lambda = \alpha/2 - \gamma$	$(-\infty, 0)$ $\{0\}$ $(0, \infty)$	only gaps every alignment scores 0 no gaps

Table 2.1: The Needleman-Wunsch scoring function can be replaced by three scoring functions which only have one parameter.

As optimality criterion in our simulation we used the Needleman-Wunsch scoring function (equation 1.1). This scoring function has three parameters: α which is the score of a match, β which is the score of a mismatch and γ the score of a gap. It turns out, however, that we can also retrieve the optimal Needleman-Wunsch alignments using a scoring function which only has one parameter which stands for the relative weight of gaps as opposed

21

to mismatches (cf. Holmes and Durbin, 1998):

$$S_\lambda(A) = -b(A) - \lambda c(A) \quad \text{with} \quad \lambda = \frac{\alpha/2 - \gamma}{\alpha - \beta}. \tag{2.3}$$

If $\alpha > \beta$, then S_λ is maximal exactly where S_{NW} is maximal. Table 2.1 summarizes the possible cases. In the following we will discuss only case I from table 2.1 in detail as not preferring matches over mismatches would be considered unreasonable in any biological application. Obviously, the optimal alignments that we get by maximizing equation 2.3 depend on the choice of λ. As we wanted to study what the optimal alignments look like with varying λ we applied the concept of parametric sequence alignment (cf. Gusfield, 1997): Since the work of Fitch and Smith (1983) it is known that the space of alignment parameters can be cut into pieces within which the optimal alignments do not change. There are also efficient algorithms (e. g. Waterman *et al.*, 1992) to find this tesselation of the parameter space, which in our case is just a partitioning of the λ-axis. Thus we can compute all optimal alignments of two sequences at a relatively low computational cost.

The complete simulation now worked like this: We generated 1000 pairs of sequences of length 100, with exactly a matches, b mismatches and $c/2$ gaps in each sequence for every triple $(a, b, c) \in M$ with $M = \{(x, y, z) \in \mathbb{N}_0^3 \mid 2x + 2y + z = 200\}$. For each pair of sequences we then performed the parametric sequence alignment with the modified Needleman-Wunsch scoring function (equation 2.3) and recorded the corresponding $\left\{ (\hat{a}_\lambda, \hat{b}_\lambda, \hat{c}_\lambda) \right\}$.

The results of this simulation are summarized in the figures 2.2 to 2.6 and in table 2.2. In order to display the results we find it convenient to define the gap ratio of an alignment A as

$$r_G(A) := \frac{c(A)}{2a(A) + 2b(A) + c(A)}. \tag{2.4}$$

If we are refering to the gap ratio of true alignments we will simply write r_G and we will denote the gap ratio of alignments that are optimal for λ as $\hat{r}_{G,\lambda}$. The gap ratio of an alignment can be interpreted as the proportion of bases that form gaps. Similarly, the proportions of an alignment A's bases which are part of mismatches and of matches can be computed as the mismatch ratio $r_S(A) := d(A)(1-r_G(A))$ and the match ratio $r_M(A) := (1-d(A))(1-r_G(A))$ respectively.

Figure 2.2 shows for every combination of the true Hamming distance per position $d \leq 0.75$ and the true gap ratio $r_G \leq 0.15$ which corresponds to an $(a, b, c) \in M$ the proportion of sequence pairs for which the true alignments were optimal for some interval of the λ-axis. One can see that only if $r_G = 0$ or if both d and r_G were very small, the existence of λ values for which the true alignments were optimal was guaranteed. With increasing r_G and with increasing d the number of sequence pairs whose true alignments were also among their optimal ones decreases until it finally equals zero. Thus, the true alignments were always suboptimal for the vast majority of the (d, r_G) pairs in our simulation.

What this means for a single point is displayed in figure 2.3. There we regard the case $(a, b, c) = (60, 30, 20)$. Thus, d is equal to $\frac{1}{3}$ and r_G equals 0.1. Considering only the non-degenerate case $\lambda > \frac{1}{2}$, we observe that none of the optimal alignments had numbers of matches, mismatches and gaps which equal the true (a, b, c) vector. They either had less gaps or less mismatches or both. We also hardly have a good chance to estimate d correctly. Only if $(\hat{a}_\lambda, \hat{b}_\lambda, \hat{c}_\lambda)$ hits the line labelled '1/3' the distance is estimated correctly.

How the observed Hamming distance per position $\hat{d}_\lambda$ varies with λ is shown in the figures 2.4 to 2.6 for some combinations of d and r_G. The shape of each of these plots results from the same cause: Large values of λ

23

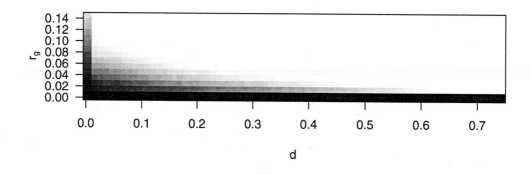

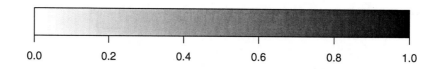

Figure 2.2: The proportion of sequence pairs for which the true alignment is one of the optimal alignments. For each point in the upper diagram 1000 pairs of sequences of length 100 with a given Hamming distance d (x-axis) and a given gap ratio r_G (y-axis) were simulated and then aligned parametrically with $\lambda > 0$ (see text). The grey levels indicate how many of these sequence pairs had optimal alignments which had the same numbers of matches, mismatches and gaps as the sequences' true alignment. Points which are white thus represent combinations of d and r_G for which the true alignment is always suboptimal no matter how one chooses λ. The lower diagram shows how to interpret the various grey levels.

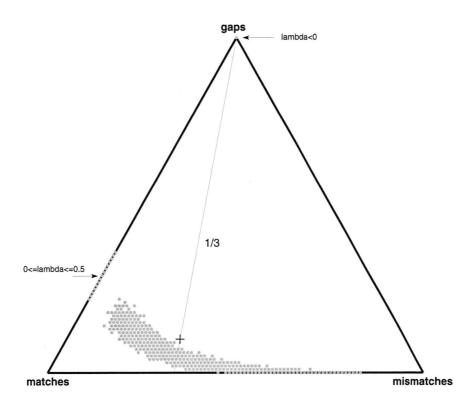

Figure 2.3: A de Finetti diagram (de Finetti, 1926) showing the empirical distribution of the number of matches, mismatches and gaps for pairs of sequences having exactly 60 matches, 30 mismatches and 10 gaps each (+) over the entire range of the alignment parameter λ. The lengths of the perpendiculars from a point in the diagram to the right, left and lower side of the triangle are proportional to the corresponding match, mismatch and gap ratios respectively. The line labelled '1/3' contains those points for which d is estimated correctly.

	$d = 0.1$	$d = 0.3$	$d = 0.5$	$d = 0.7$
$r_G = 0.0$	∞ / 1	∞ / 1	∞ / 1	∞ / 1
$r_G = 0.1$	1.08 / 1.33	2.17 / 1.21	5.77 / 1.28	125 / 1.05
$r_G = 0.2$	0.769 / 1.28	1.49 / 1.10	3.77 / 1.24	78.7 / 1.06

Table 2.2: The values of $\hat{\lambda}$ (left of the slashes) which Holmes and Durbin (1998) predict to optimize the alignment fidelity given that the true alignment's gap ratio is equal to r_G and that the true Hamming distance is equal to d and the corresponding mean of $\hat{d}_\lambda/d$ (to the right of the slashes).

will force the sequences to align without gaps. While this gives rise to the possibility to get perfect estimates in the case that the true alignments did not contain any gaps (figure 2.4), a large λ leads to an overestimation of the sequences' distance when the true alignments have gaps (figures 2.5 and 2.6). In the latter case, the pairwise alignment algorithm will falsely align positions which are not related. These alignment positions will then have a mean Hamming distance of $\frac{3}{4}$ given that all bases occur at the same frequency. Therefore, the overestimation increases with the number of gaps in the true alignments (compare the corresponding diagrams in figure 2.5 and 2.6) and decreases with the true Hamming distance per position. The most striking result of this simulation is the fact that it is almost impossible to get reliable estimates of the Hamming distance when d is small and the true alignments include gaps.

Holmes and Durbin (1998) give a formula to calculate the relative gap weight $\hat{\lambda}$ which maximizes the alignments fidelity from the probabilities of insertions, deletions and substitutions. Although they used a different way to simulate their sequences, their method produces, in the expectation, the same sequence pairs as ours. Hence, we computed $\hat{\lambda}$ and the corresponding

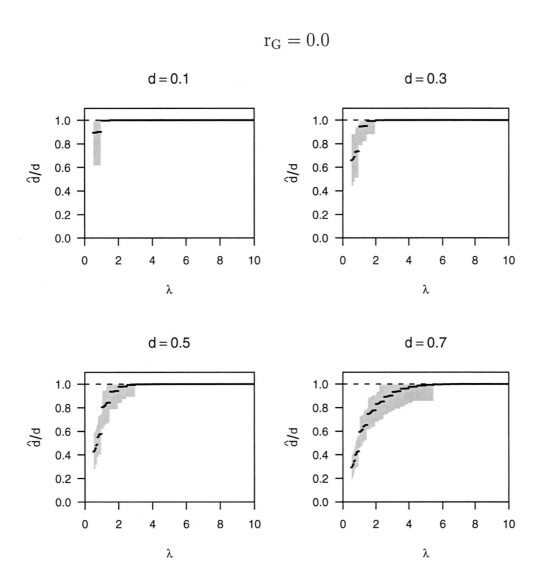

Figure 2.4: The ratio of the estimated and the true Hamming distance plotted against the alignment parameter λ for sequences of length 100 when the gap ratio r_G is equal to 0.0 and when d is equal to the value indicated on top of each diagram. The black points represent the means of 1000 simulations and the grey area is bounded by the 2.5% and 97.5% quantiles. Along the dashed line, $\hat{d}_\lambda$ equals d.

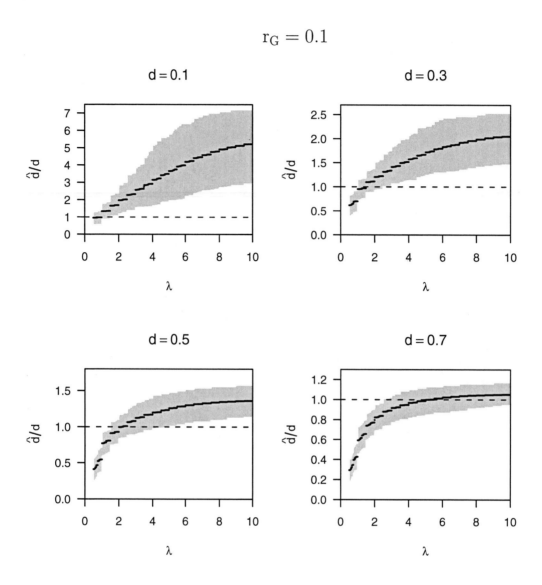

Figure 2.5: The ratio of the estimated and the true Hamming distance plotted against the alignment parameter λ for sequences of length 100 when the gap ratio r_G is equal to 0.1 and when d is equal to the value indicated on top of each diagram. The black points represent the means of 1000 simulations and the grey area is bounded by the 2.5% and 97.5% quantiles. Along the dashed line, $\hat{d}_\lambda$ equals d. Note the different scalings of the y-axes.

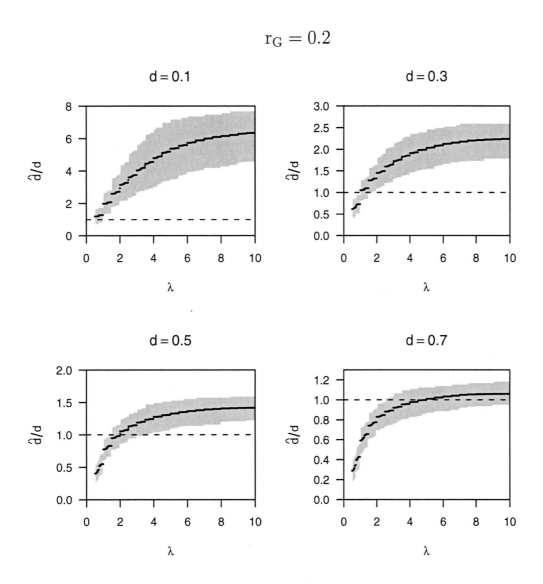

Figure 2.6: The ratio of the estimated and the true Hamming distance plotted against the alignment parameter λ for sequences of length 100 when the gap ratio r_G is equal to 0.2 and when d is equal to the value indicated on top of each diagram. The black points represent the means of 1000 simulations and the grey area is bounded by the 2.5% and 97.5% quantiles. Along the dashed line, $\hat{d}_\lambda$ equals d. Note the different scalings of the y-axes.

means of $\hat{d}_\lambda$ for the (d, r_G) pairs from figures 2.4 to 2.6. As one can see in table 2.2 even $\hat{\lambda}$ does not yield correct distances for $r_G > 0.0$. Note also that for any pair of sequences whose true alignment is unknown it is impossible to know the value of $\hat{\lambda}$ unless we have a full-grown insertion-deletion model, in which case we have more sophisticated techniques at hand than optimal Needleman-Wunsch alignment (see chapter 4).

2.3 A least squares approach to distance estimation

Since the performance of the naïve estimator $\hat{d}_\lambda$ is rather poor, we studied a least squares approach to estimate evolutionary distances between pairs of sequences in a more reliable manner. The goal is to include information of the various $(\hat{a}_\lambda, \hat{b}_\lambda, \hat{c}_\lambda)$ statistics for a pair of sequences. To do this we introduce the variable $X(A) := (r_M(A), r_S(A), r_G(A))$, which is the triple consisting of the match, mismatch and gap ratio of alignment A. Again, we will write X instead of $X(\mathcal{A}_{true})$ and $\hat{X}_\lambda$ instead of $X(A_{opt,\lambda})$. With parametric sequence alignment, a given pair of sequences, whose true alignments are unknown, produces a collection $\left\{\hat{X}_\lambda\right\}_{\lambda \in \Lambda}$ of match, mismatch and gap patterns, where $\lambda \in \Lambda$ is the center of an interval that gives rise to the same maximal alignment and estimate of the unknown X. The problem is to infer the unknown match, mismatch and gap ratios from $\left\{\hat{X}_\lambda\right\}_{\lambda \in \Lambda}$. We suggest the following least squares approach:

$$\hat{X}_{LS} := \underset{X}{\mathrm{argmin}} \left\{ \sum_{\lambda \in \Lambda} \left(\hat{X}_\lambda - \mathbb{E}(\hat{X}_\lambda | X) \right)^2 \right\},$$

where $\mathbb{E}(\hat{X}_\lambda | X)$ is the expectation of $\hat{X}_\lambda$ given X. While $\left\{\hat{X}_\lambda\right\}_{\lambda \in \Lambda}$ can be estimated quite quickly, we have to compute $\left\{\mathbb{E}(\hat{X}_\lambda | X)\right\}_{\lambda \in \Lambda}$ by extensive simulation employing the parametric alignment scheme. For a given X we produce 1000 pairs of sequences (see section 2.2), which are then aligned parametrically. Each parametric alignment induces a partition of the λ-axis. The intersection of all these partitions is again a set of disjoint intervals. For each interval one thousand X vectors are observed. Thus we can estimate $\left\{\mathbb{E}(\hat{X}_\lambda|X)\right\}_{\lambda\in\Lambda}$ as the average of 1000 simulated pairs of sequences.[2]

Figure 2.7 displays the averages of the $\hat{X}_{LS}$ estimates taken over 100 simulations for sequences of length 100. Even for sequences this short the least squares approach allows quite good estimates of mismatch and gap ratios when these are not too big, and thus also of d. For larger mismatch and gap ratios, however, the estimates all lie near the center of the diagram. At first sight, this bias towards the center looks strange, as it means that for some true mismatch and gap ratios the values after aligning are in the mean closer to the mean values of other true mismatch and gap ratios than they are to their own ones. Yet, there are two possible explanations for this phenomenon: Firstly, we do not account for the variance of the distributions. This means we are being unfair to the broader distributions. Secondly, for true values at the border of the diagram wrong estimates will always lie closer to the center.

The quite good performance of the least squares method for sequences of length 500 is shown in the figures 2.8 and 2.9. What is remarkable are the very small standard deviations of the estimates of d when the true alignment did not not contain gaps. Yet, this method, too, leads to an overestimation for small values of d when $r_G = 0.1$.

---

[2]For the case of distance estimation from ungapped local alignments, Agarwal and States (1996) developed a similar method which relies also on parametric alignment but uses posterior probabilities instead of least squares. In our case the least squares approach is however much easier to handle.

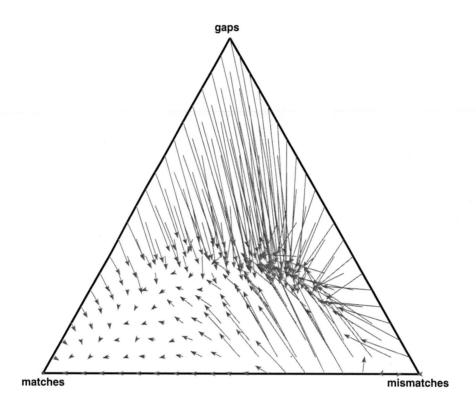

Figure 2.7: The mismatch and gap ratios estimated with the least squares approach (tips of the arrows) for several true values of r_S and r_G (the arrows' starting points). The sequence length was 100. The alignments were drawn from the ones produced in 2.2.

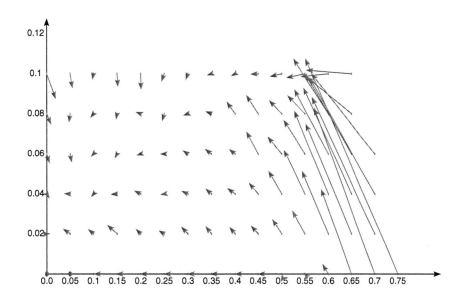

Figure 2.8: The performance of our method for sequences of length 500. One can see the mean of the estimates (taken over 100 simulations).

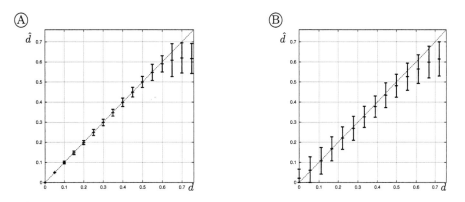

Figure 2.9: The performance of our method for sequences of length 500. One can see the average of the estimates (based on 100 simulations for each true distance) of the number of differences per position plotted against the true values. The vertical bars display plus/minus two times the standard deviation. Ⓐ: $r_G = 0$, Ⓑ: $r_G = 0.1$

33

2.4 Discussion

So far, the bias introduced into the estimation of pairwise differences when aligning two sequences has not been recognized and therefore has not been systematically studied. We have shown that the evolutionary distance between sequences cannot be estimated reliably if the sequences were subjected to an alignment procedure. This statement holds true even if the entire space of possible alignment parameters is studied. One might argue that sequences of length 100 are too short to make valuable assertions, yet given the work of Hwa and Lässig (1996) the problem should even get worse for longer sequences, since the fluctuations of optimal alignment paths increase with sequence lengths. Note also that the problems outlined in this chapter have severe implications for the reconstruction of phylogenetic trees. If we tried to reconstruct a phylogeny for a set of sequences using distance estimates that we can obtain from their optimal pairwise alignments, we would hardly get the correct tree. The more so as alignment parameters which might lead to good estimates for one sequence pair of a data set, might yield biased estimates for another one (see figures 2.4 to 2.6). Although we normally use multiple instead of pairwise alignments when doing phylogenetic inference, many multiple alignment programs (e. g. Thompson *et al.*, 1994) start with the computation of optimal pairwise alignments from which pairwise distances are estimated.

We have shown that, using a least squares approach, one can improve the estimation of evolutionary distance and also of the gap frequencies, for a large region of the parameter space. Moreover, our approach is an efficient way of utilizing the information provided by parametric sequence alignment. Yet, we have to rely on empirical distributions of the numbers of matches, mismatches and gaps given the respective numbers in the true alignments and given the

34

alignment parameter λ. In order to apply our method to scenarios of sequence evolution different from the one we used here, for instance unequal base frequencies or rate heterogeneity, we would therefore have to redo at least some of our simulations to fathom these empirical distributions. The only way to circumvent this would be an analytic result. Considering, however, the enormous mathematical difficulties that one already encounters when trying to distinguish between the local alignment scores of related sequences and of random ones (cf. Siegmund and Yakir, 2000), a parameterized distribution of some summary statistics of optimal pairwise sequence alignments which takes into account the parameters of an evolutionary process and the alignment parameters seems presently out of reach.

Chapter 3

Reconstructing trees from multiple alignments

3.1 Introduction

As they are the prerequisite for phylogenetic inference, the various tree reconstruction methods and programs have been studied in great detail. Analytical examinations have shown that while some methods like for instance maximum parsimony (Farris *et al.*, 1970; Fitch, 1971) may be inconsistent[1] (Felsenstein, 1978b; Zharkikh and Li, 1993), others like maximum likelihood or minimum evolution will reconstruct the correct topology given that the sequences are long enough and given that the underlying assumptions are not violated (cf. Rogers, 1997; Huelsenbeck and Hillis, 1993). Extensive simulation studies (e. g. Huelsenbeck, 1995; Schöniger and von Haeseler, 1995a) have been carried out to assess the robustness of the commonly used tree-building methods against violations of their assumptions and to get an idea

---

[1] An estimation procedure is called inconsistent if it fails to yield the correct value even if provided with infinite data

how many sites are needed for an accurate inference. Yet, all these investigations presumed the true alignments of the sequences to be known. Studies of multiple alignment algorithms, on the other hand, usually focus on the methods' ability to find the optimum of some scoring function (e. g. Gonnet *et al.*, 2000) or on their power to detect sequence motives that are common to a given protein family (McClure *et al.*, 1994; Briffeuil *et al.*, 1998; Thompson *et al.*, 1999) and are not interested in phylogenetic inference. Thus, the effects that alignment errors may have on the reconstructed trees are normally ignored. Morrison and Ellis (1997) have shown, however, that this is not only an academic question but that phylogeny estimation may indeed be influenced severely by sequence alignment. They aligned a data set of 18S rDNA sequences with six different alignment methods. From the obtained alignments they then estimated the sequences' phylogeny using three tree reconstruction algorithms. They found that the different alignment methods caused more variation in the reconstructed phylogenies than did the usage of different tree reconstruction methods.

In this chapter, we describe three simulations that investigate the success rate of phylogenetic inference including also the alignment procedure. As these simulations were very time-consuming we only studied two multiple alignment methods: the widespread progressive alignment program CLUSTALW (Thompson *et al.*, 1994), and PRRN (Gotoh, 1996) which is an implementation of an iterative refinement method (see chapter 1) that was reported to perform better than CLUSTALW in some tests (Gotoh, 1996; Thompson *et al.*, 1999). The various parameters of the two programs remained unchanged except of the gap penalties which were modified during the simulations I and II. Note that both programs use affine gap penalties, i. e. the opening of a new gap is penalized differently than the extension of an existing gap. Note also that unlike the gap score introduced in section 1.1 these gap penalties are subtracted from an alignment's score and not added to it.

3.2 Simulation I: The four-taxon case

In our first simulation we used the model tree depicted in figure 3.1. This four-taxon tree has two parameters: the two-edge rate s_2 which specifies the expected number of substitutions that occur on the branches leading to the sequences 1 and 3 and the three-edge rate s_3 which is the length of the tree's central branch as well as the length of the branches leading to the sequences 2 and 4. Felsenstein (1978b) has shown that if sequences evolve according to this tree and if s_2 is large as compared to s_3, maximum parsimony will be inconsistent. This is also the tree that was used in the studies of Huelsenbeck and Hillis (1993) and Schöniger and von Haeseler (1995a) to demonstrate that many other methods also fail to be consistent for certain values of s_2 and s_3, especially if the distances between the sequences are underestimated or if the evolution of the sequences is not modelled appropriately.

We adopted the simulation scheme of Huelsenbeck and Hillis (1993) and simulated for each pair (s_2, s_3) whose corresponding expected numbers of sequence differences d_2 and d_3 lie in the set $\{0.01, 0.03, \ldots, 0.73, 0.75 - 10^{-6}\}$ 100 artificial data sets of DNA sequences with the help of the program SEQGEN (Rambaut and Grassly, 1997). As substitution model we used the one introduced by Jukes and Cantor (1969) and the sequence length was set equal to 200, 1000 and 5000 in three independent runs of the entire simulation. Each simulated data set was then aligned with CLUSTALW (version 1.8) and with PRRN (version 2.5.2). CLUSTALW used its default gap-opening

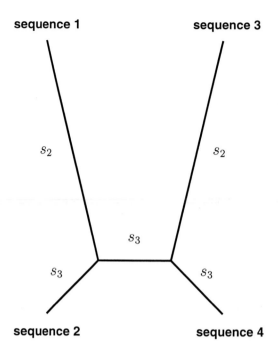

Figure 3.1: The four-taxon tree used in simulation I. The branch leading to sequence 1 has the same length (s_2) in terms of expected substitutions per position as the branch leading to sequence 3. The three other branches have length s_3.

and gap-extension penalties of 10.0 and 0.1 respectively. PRRN's gap-opening penalty was set to 10 and its gap-extension penalty to 1. This was as close as we could get to CLUSTALW's settings, since PRRN only accepts integer values. We then removed those alignment positions which contained gaps and estimated the pairwise Jukes-Cantor corrected distances separately for each of the two alignments as well as for the original data set. Finally, we reconstructed neighbor-joining trees from these pairwise distances.

Table 3.1 shows how often we were successful in reconstructing the correct phylogeny. The higher the success rate, the darker are the points in the depicted diagrams. While the success rates increase with increasing sequence length when the trees are estimated directly from the simulated data sets without an additional alignment step, the usage of the two alignment programs leads to very low success rates in some areas of the grid even if the sequences are quite long. Both programs result in a success rate of virtually 0 when d_2 is above and d_3 below certain thresholds. The almost rectangular shape of this white area might look peculiar when one thinks of e. g. maximum parsimony's zone of inconsistency. Yet, similar shapes have been observed by Huelsenbeck (1995) when the pairwise distances were overestimated. Additionally to the upper left corner of the diagrams, tree reconstruction from CLUSTALW alignments exhibits a drastically reduced success rate when d_3 gets too big. We cannot say, however, whether the usage of multiple alignment programs really creates inconsistencies or if the light areas in our plots finally would become black if we increased the length of the data sets even further. Yet, as the boundaries between the dark and the light areas in the plots get sharper with increasing sequence length and as they do not shift, the latter does not seem very plausible.

3.3   Simulation II: Varying gap penalties

The sequences in the last section were simulated without any insertions or deletions. Therefore, if only the gap penalties had been big enough, the two multiple alignment programs should have been able to retrieve the correct alignment and thus would not have had a negative effect on the success rate of tree reconstruction. In order to see whether it is always possible to pick the gap penalties in a way that tree reconstruction is not affected by alignment errors, we simulated four-taxon data sets which did not only contain substitutions but also insertions and deletions and we varied the gap-opening and the gap-extension penalties of CLUSTALW and PRRN. Like in the previous

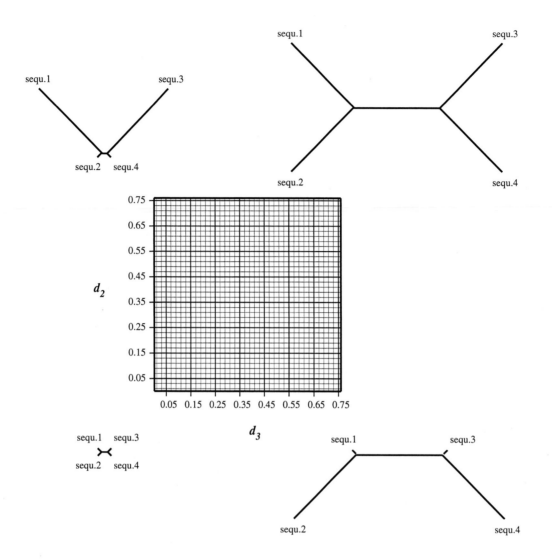

Figure 3.2: This figure illustrates how to read the plots in table 3.1. The length of the interior branch and of the branches leading to the sequences 2 and 4 (s_3, see figure 3.1) increases along the horizontal axis. The length of the two other branches (s_2) grows in vertical direction. d_2 is the expected Hamming distance per position that corresponds to the substitution rate s_2 under the Jukes-Cantor model (Jukes and Cantor, 1969). Accordingly, d_3 is the expected Hamming distance per position that follows from substitution rate s_3. The depicted trees sketch the situation at the corners of the diagram.

42

	$l = 200$	$l = 1000$	$l = 5000$
alignment known			
CLUSTALW			
PRRN			

Table 3.1: The success rate of neighbor-joining for data sets that have evolved along the tree depicted in figure 3.1 without insertions or deletions. The columns of the table correspond to simulated data sets of 200 (left), 1000 (middle) and 5000 positions (right). The rows show the results for the tree reconstruction based upon the true alignments (top), CLUSTALW alignments (middle) and PRRN alignments (bottom). See the text for the parameter settings. Each diagram is organised in the way shown in figure 3.2. 100 data sets were simulated for each node of the grid $\{0.01, 0.03, \ldots, 0.73, 0.75 - 10^{-6}\} \times \{0.01, 0.03, \ldots, 0.73, 0.75 - 10^{-6}\}$. The grey levels are proportional to the number of correctly reconstructed topologies. Thus, black points correspond to a success rate of 100%, and white points mean that the topology was never reconstructed correctly.

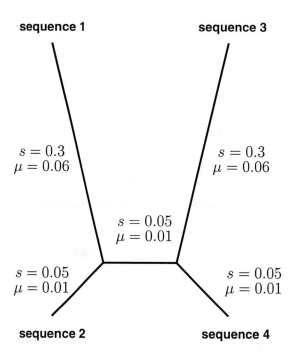

Figure 3.3: The settings for the substitution rate s and the deletion rate μ during simulation II.

simulation the substitutions occurred according to the Jukes-Cantor model. Insertions and deletions were simulated with the TKF1 model (see chapter 4). See figure 3.3 for the substitution and deletion rates, that were used in this simulation. 100 of these data sets with expected sequence length 1000 were simulated and then aligned with CLUSTALW (version 1.8) and with PRRN (version 2.5.2). The programs' gap-opening penalties were set to integer values between 1 and 20 and the gap-extension penalties were set to integer values between 0 and the gap-opening penalty used. The phylogenies were inferred in the same way as in section 3.2.

The results of this simulation are summarized in figure 3.4. For both alignment programs, the choice of the gap penalties strongly affects our ca-

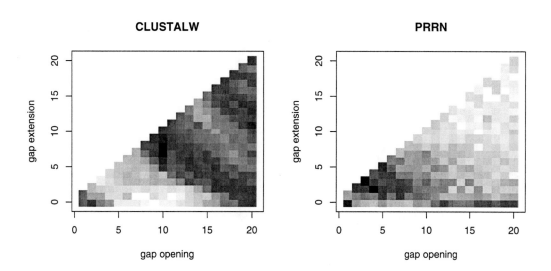

Figure 3.4: The success rate of neighbor-joining for data sets that evolved along the tree shown in figure 3.3 when the sequences are aligned with CLUSTALW (left diagram) or with PRRN (right diagram). The horizontal axes show the gap-opening penalties of the programs, and the gap-extension penalties are drawn along the vertical axes. In order to increase the contrast of the plots the grey levels of the points are proportional to the difference of the success rate and the minimum success rate of the respective plot normalized by the difference of the maximum and the minimum success rate. Therefore, the grey levels of the two diagrams must not be compared. See also table 3.2.

	minimum success rate	maximum success rate
CLUSTALW	0.32 (8,1),(11,1)	0.62 (10,7)
PRRN	0.10 (17,15),(17,16),(18,15)	0.62 (4,2)

Table 3.2: The minimum and the maximum success rate of neighbor-joining observed in simulation II. The values in brackets are the combinations of gap-opening and gap-extension penalties for which the minimal and the maximal rates of success are obtained.

pability to retrieve the correct topology. Furthermore, gap penalties which lead to a relatively good performance of one alignment program may cause the other one to perform badly. Table 3.2 shows that even for an optimal choice of gap penalties we get less than two thirds of the topologies correctly. For the worst choice of gap penalties, on the other hand, trying to reconstruct the four-taxon trees from the multiple alignments does not work better than guessing in the case of CLUSTALW or even significantly worse in the case of PRRN.

3.4   Simulation III: Influence on larger trees

In order to get an idea in how far the reconstruction of larger phylogenetic trees may be affected by alignment errors we applied a simulation scheme which has been used in several studies (Saitou and Nei, 1987; Schöniger and von Haeseler, 1993; Strimmer, 1997) to evaluate the performance of various tree reconstruction methods, but added an alignment step between the simulation of the sequences and the reconstruction of their phylogeny. For each of the three tree topologies depicted in figure 3.5 and for each of the four combinations of branch lengths given in figure 3.5①D we generated 100 artificial data sets of DNA sequences with the expected length of the sequences being 1000. Substitutions occurred following Kimura's two-parameter model (Kimura, 1980) with the transition-transversion parameter τ (see table A.1 in appendix A.1) equalling 4 and with an overall substitution rate of 0.01 substitutions per position and per unit of time. Additionally to this substitution process, short subsequences were inserted or deleted according to the TKF2 model (Thorne *et al.*, 1992, see also chapter 5) with an insertion rate of 0.002. The length of insertions and deletions was distributed geometrically with 10 being the expected indel length. For each of these data sets we recorded its true alignment as well as the unaligned sequences. These were then aligned with CLUSTALW (version 1.8) and PRRN (version 2.5.2), both programs using their default settings. Thus, we got for every simulated data set three multiple alignments: A CLUSTALW alignment, a PRRN alignment and the true alignment that resulted from simulating the sequences. After removing the positions which contained gaps, we reconstructed a quartet-puzzling tree (Strimmer and von Haeseler, 1996) for each of these alignments using version 4.0.2 of the PUZZLE program. The model that was assumed by PUZZLE was the one introduced by Tamura and Nei (1993) which is equivalent to Kimura's two-parameter model if its pyrimidine-purine-transition parameter κ (see table A.1) is set to 1 and if all base frequencies are equal. Hence there were four trees for every data set: T_{true} the tree that was used to create the data, T_{rec} the quartet-puzzling tree that was reconstructed from the true alignment, T_C the tree that was estimated based upon the CLUSTALW alignment and T_P the tree that we got from the PRRN alignment.

The figures 3.6 to 3.11 summarize the results of this simulation. Figure 3.6 shows that, no matter which tree topology we look at, the proportion of posi-

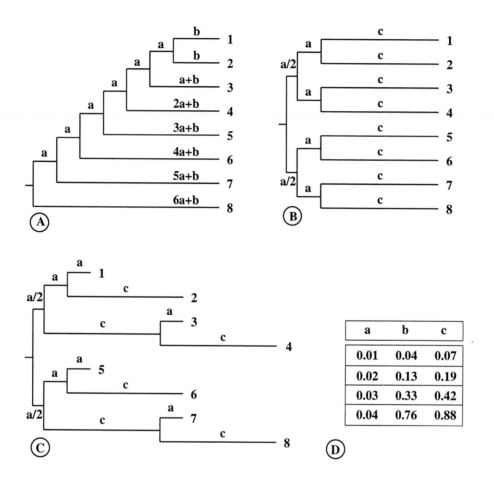

Figure 3.5: The three tree topologies of simulation III. Ⓐ: an asymmetric tree with clock-like evolution. Ⓑ: a symmetric tree with clock-like evolution. Ⓒ: a symmetric tree with non-clock-like evolution. Ⓓ: the four combinations of branch lengths (in terms of expected substitutions per position) used during the simulation.

48

CLUSTALW 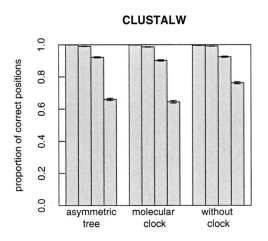   **PRRN**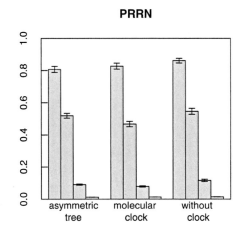

Figure 3.6: The proportion of positions in the true alignments that are also present in the alignments found by CLUSTALW (left side of the figure) and by PRRN (right side). The height of the bars shows the mean of 100 simulations, the error bars represent the standard error. The labelling of the bottom line corresponds to the three tree topologies from figure 3.5. For each topology the branch lengths that were used for the simulation (see figure 3.5 ⓓ) increase from left to right.

tions of the true alignments that were correctly recovered by the two multiple alignment programs decreases with increasing sequence divergence. This is in good agreement with the findings of Gotoh (1996), Briffeuil *et al.* (1998) and Thompson *et al.* (1999). However, unlike the results in Gotoh (1996) and unlike simulation I, PRRN performed significantly worse than CLUSTALW in this simulation. This decrease in the alignments' fidelity is accompanied by an increase in the partition distance between the true topologies and the ones which can be reconstructed from these alignments, as is shown in figure 3.7. The partition distance $\Delta(T_1, T_2)$ of two trees T_1 and T_2 is simply the number of clusters which are only present in one of them but not in both and is a convenient measure for the dissimilarity of the two trees (Robinson

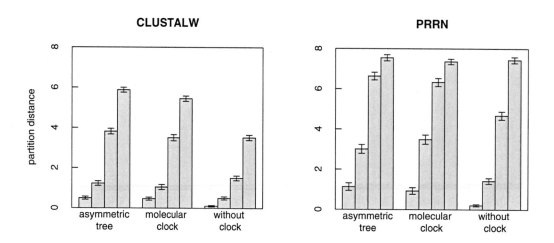

Figure 3.7: The mean partition distance between the true tree and the puzzle trees that were reconstructed based upon the alignments found by CLUSTALW (left side of the figure) and by PRRN (right side). The arrangement of the bars is the same as in figure 3.6.

and Foulds, 1981). In order to see whether the increase in partition distance with increasing branch lengths is due to alignment errors we computed for every data set the difference between $\Delta(T_C, T_{true})$ and $\Delta(T_{rec}, T_{true})$ and the one between $\Delta(T_P, T_{true})$ and $\Delta(T_{rec}, T_{true})$. As can be seen in figure 3.8 the trees that were reconstructed from the CLUSTALW alignments are almost as close to the true trees as are the ones which were reconstructed from the true alignments. The trees that were estimated from the PRRN alignments are, however, significantly worse.

As quartet-puzzling trees are consensus trees, they need not be bifurcating but may contain unresolved nodes. Thus, we can address the question whether aligning the sequences is just blurring the data's phylogenetic signal or whether multiple alignment methods may force a wrong tree structure upon the data. To do this we define the degree of unresolvedness $u(T)$ of a

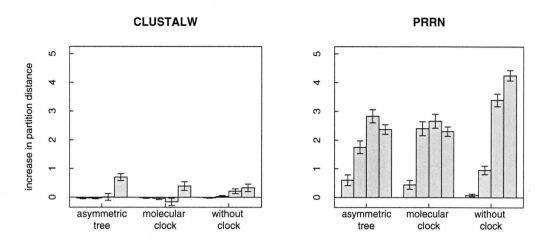

Figure 3.8: The mean increase in partition distance due to alignment errors. The arrangement of the bars is the same as in figure 3.6.

tree T with n leaves as $1 - e_T/(n-3)$, where e_T is the number of T's internal edges and $n - 3$ is the number of internal edges of a completely resolved n-taxon tree. Figure 3.9 shows that the unresolvedness of the trees that were reconstructed from the CLUSTALW and PRRN alignments increases monotonically with the branch lengths of the true trees. If we do, however, have a look at the differences $u(T_C) - u(T_{rec})$ and $u(T_P) - u(T_{rec})$ (see figure 3.10), we have to notice that the trees T_C and T_P are in the mean better resolved than the ones we estimated from the true alignments. Hence, the two multiple alignment programs squeeze the data in a treelike shape. Unfortunately, it tends to be the wrong one (see figure 3.8).

That the alignment programs did not only influence the shape of the reconstructed phylogenies but also the estimates of the transition-transversion parameter τ, can be seen in figure 3.11. While CLUSTALW does not lead to a biased estimation of τ, PRRN alignments yield smaller and smaller estimates the more the sequences are diverged. This difference in the behaviour

51

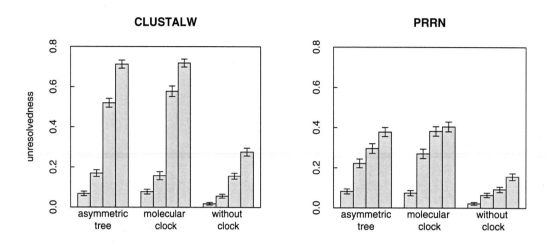

Figure 3.9: The mean degree of unresolvedness (see text) of the reconstructed trees. The arrangement of the bars is the same as in figure 3.6.

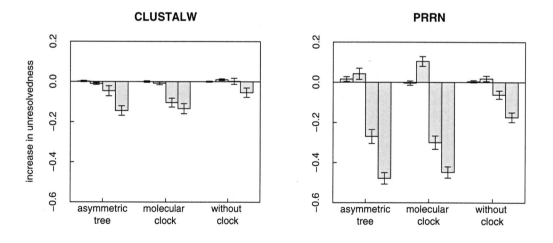

Figure 3.10: The mean increase in unresolvedness (see text) due to alignment errors. The arrangement of the bars is the same as in figure 3.6.

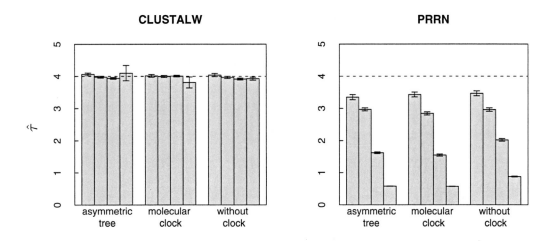

Figure 3.11: The means of the estimates $\hat{\tau}$ of the transition-transversion parameter when PUZZLE is provided with CLUSTALW alignments (left) or PRRN alignments (right). The dashed line corresponds to the value that was used for the simulation of the sequences. The arrangement of the bars is the same as in figure 3.6.

of the two alignment programs is probably caused by their different treatment of mismatches: CLUSTALW penalizes transitions less than it penalizes transversions, PRRN, however, does not distinguish between these two types of mismatches. Hence, PRRN is implicitly assuming the same substitution process as Felsenstein (1981) and pushes the data towards this simpler substitution model. Note also that the estimates of the pyrimidine-purine transition parameter κ were always close to the true value 1.0 no matter which alignment method was used (data not shown).

3.5 Discussion

The simulations described in this chapter exemplify some of the problems that one has to be aware of when using multiple sequence alignments in phylogenetic research: Multiple alignment programs make implicit assumptions about the plausibilities of mismatches and gaps. By projecting these views on the data, they may be the cause of erroneous conclusions about the relationship of the sequences under study (see figure 3.8) or of biased parameter estimates (see figure 3.11) and may even increase our confidence in the wrong phylogeny (see figures 3.8 and 3.10). Thus, although we have accurate tree reconstruction methods at hand, the necessity to align the sequences will in some cases prevent us from getting the correct topology (see table 3.1).

One should note, however, that both alignment programs performed reasonably well when the branch lengths of the model trees were not too extreme. Therefore, this chapter's results should not be seen as a criticism of multiple alignment programs, but rather as a pleading for the development of methods which allow to assess the reliability of these programs' output.

Chapter 4

Statistical pairwise alignment

4.1 Introduction

In chapter 2 we described the problems that arise from the traditional way
to study a pair of biological sequences: As we are only modelling the substitution process but not the processes which create insertions and deletions,
we have to find an optimal alignment of the sequences before any further
analysis. Yet, our speculations about the properties of the evolutionary process which has produced the sequences will influence which alignment we will
consider to be the best, and this alignment will in turn affect our conclusions
concerning the sequences' evolution. If we possessed, however, a reasonable
and computationally tractable way to model the creation of insertions and
deletions, we would not only be able to find the most probable alignment
of two sequences or to get maximum likelihood estimates of the model parameters, but we could also address questions like the following: What is
the posterior distribution of the model parameters? Which parts of an alignment are relatively certain and which ones are more questionable? What do
typical alignments look like? How do the typical alignments depend on the

parameters of the model?

Several attempts have been made to specify such a model: While some (e. g. McGuire *et al.*, 2001) try to model insertions and deletions in the same way as they model substitutions, others (e. g. Holmes and Durbin, 1998) construct hidden Markov models (HMMs) which create two sequences from left to right, but need not correspond to an evolutionary process in time. Following the work of Bishop and Thompson (1986), a number of models has been described which try to provide both a reasonable insertion-deletion dynamics and computational tractability (Thorne *et al.*, 1991, 1992; Miklós and Toroczkai, 2001; Metzler, 2003). Yet, only the one introduced by Thorne *et al.* (1991), which we refer to as the TKF1 model, has gained greater attention.

In the following we are going to give a brief and informal review of the TKF1 model (section 4.2) and then we are going to describe one of its possible applications, namely a method which jointly samples pairwise alignments and mutation parameters (section 4.3).

4.2   The TKF1 insertion-deletion model

The TKF1 model contains three parameters: the substitution rate s, the insertion rate λ, and the deletion rate μ. Each site is independently deleted with rate μ, and hit by a substitution with rate s. Insertions occur between any two sites or at the ends of the sequence at rate λ. When a substitution or an insertion occurs, the new base is drawn randomly from $\{A, G, C, T\}$ according to a probability distribution $(\pi_A, \pi_G, \pi_C, \pi_T)$ [1].

Given a sequence of length n, there are n sites which are candidates for

---

[1] The last sentence implies that a nucleotide may be substituted by a nucleotide of the same type. During this chapter we will stick to this rather odd definition of the word "substitution" as this is how the TKF1 model was originally formulated.

deletion and $n + 1$ positions where a nucleotide can be inserted. Therefore, the length of the sequence grows with net rate $\lambda(n+1) - \mu n$. In order to avoid a drift towards longer and longer sequences it is assumed that $\mu > \lambda$, which then ensures the existence of a stationary distribution of the sequence evolution process. At equilibrium, the number of possible insertion sites, i. e. the length of the sequence plus 1, is geometrically distributed with expectation $\frac{\mu}{\mu - \lambda}$ and the bases are independently distributed according to $(\pi_A, \pi_G, \pi_C, \pi_T)$. This time stationary process is *reversible*: the probability of starting with an ancestral sequence s_1 and arriving after a time t at an offspring sequence s_2 is unaltered when s_1 and s_2 are interchanged.

In general, certain nucleotides of the ancestral sequence will be conserved in the offspring sequence; other nucleotides will appear only in the ancestral or only in the offspring sequence. Assume for instance that the ancestral sequence is ACGTC; it may happen that A is substituted by a C, the G is deleted and an A is inserted between the T and the C. Then, with the ancestral sequence on top, this results in the alignment

$$
\begin{array}{c}
\texttt{ACGT\_C} \\
\texttt{CC\_TAC}
\end{array}
\qquad (*)
$$

Other sequence histories lead to ambiguities in the alignment which must be resolved by convention. For instance, starting with ACG, suppose C is deleted and then T is inserted between A and G. There are two possibilities for the resulting alignment:

$$
\text{alignment 1:}
\begin{array}{c}
\texttt{AC\_G} \\
\texttt{A\_TG}
\end{array}
\qquad \text{or alignment 2:}
\begin{array}{c}
\texttt{A\_CG} \\
\texttt{AT\_G}
\end{array}
$$

Thorne *et al.* (1991) make the convention that insertions happen to the right of a nucleotide rather than between nucleotides. Thus in the example, T must be the offspring nucleotide of A and alignment 2 is the appropriate one. This would also be the case if T had been inserted between the A and the C before the deletion of the C, whereas alignment 1 would correspond to a history in which T was inserted between the C and the G and the C was deleted after that. This makes alignment 2 more probable than alignment 1. If the bottom sequence would be the ancestral one, it would be the other way round: alignment 1 would be more probable than alignment 2. Thus the alignment convention destroys the time reversibility. But this affects only the order of gaps in the two sequences and not the site homology, which is the primary objective of alignment problems. Therefore, if two sequences are given which have descended from some common ancestor, we may assume that the second sequence has evolved from the first.

Now suppose that for given parameters s, λ, and μ, a sequence is drawn from the equilibrium distribution and evolves for a time span t according to the TKF1 model. How can we then compute the probability, that, say, the alignment (*) is the true one? The first observation is that we can separate the process into two independent components: the insertion/deletion process and the process of substitutions. First we calculate the probability of the *bare alignment*

$$\begin{matrix} \texttt{BBBB\_B} \\ \texttt{BB\_BBB,} \end{matrix} \qquad (**)$$

ignoring the base types (B just stands for "Base"). Then we compute the conditional probability for the alignment (*) given the bare alignment (**), which is easy, because all $\genfrac{}{}{0pt}{}{\texttt{B}}{\texttt{B}}$, $\genfrac{}{}{0pt}{}{\texttt{B}}{\texttt{\_}}$, and $\genfrac{}{}{0pt}{}{\texttt{\_}}{\texttt{B}}$ take their base types independently of each other. For the pair $\genfrac{}{}{0pt}{}{\texttt{A}}{\texttt{C}}$ and time t, this is the product of the probability of A, the probability that at least one substitution occurs in a time span t, and the probability that the new base is C: $\pi_\texttt{A} \cdot (1 - e^{-st}) \cdot \pi_\texttt{C}$. For the pair $\genfrac{}{}{0pt}{}{\texttt{C}}{\texttt{C}}$ we have two possibilities: no substitution or one or more substitutions with the last one being C: $\pi_\texttt{C} \cdot [e^{-st} + (1 - e^{-st}) \cdot \pi_\texttt{C}]$. The probabilities for $\genfrac{}{}{0pt}{}{\texttt{G}}{\texttt{\_}}$ (given $\genfrac{}{}{0pt}{}{\texttt{B}}{\texttt{\_}}$) or $\genfrac{}{}{0pt}{}{\texttt{\_}}{\texttt{A}}$ (given $\genfrac{}{}{0pt}{}{\texttt{\_}}{\texttt{B}}$) are π_G and π_A. In the sequel we rescale the time so that

58

the time elapsed between the sequences is $t = 1$.

The computation of the probability of the bare alignment (**) is slightly more difficult, because the positions are not independent. But as a consequence of the TKF1 alignment convention, the bare alignment is generated by a Markov chain on the states $\texttt{Start}$, $\frac{B}{B}$, $\frac{B}{\_}$, $\frac{\_}{B}$ and $\texttt{End}$. Therefore, we can compute the probability of a bare alignment by stepping through it from left to right and iteratively multiplying the result with the transition probability from the preceding state to the current one. For given states $x, y \in \{\texttt{Start}, \frac{B}{B}, \frac{B}{\_}, \frac{\_}{B}, \texttt{End}\}$, we denote the probability for the next state being y, given that the current state is x, by $P_{\lambda,\mu}(x \to y)$. For the calculation of the transition probabilities (see table 4.1) we refer to Holmes and Bruno (2001).

	$y = \frac{B}{B}$	$y = \frac{B}{\_}$	$y = \frac{\_}{B}$	$y = \texttt{End}$
$x = \texttt{Start}$	$(1-\beta(t))\frac{\lambda}{\mu}\alpha(t)$	$(1-\beta(t))\frac{\lambda}{\mu}(1-\alpha(t))$	$\beta(t)$	$(1-\frac{\lambda}{\mu})(1-\beta(t))$
$x = \frac{B}{B}$	$(1-\beta(t))\frac{\lambda}{\mu}\alpha(t)$	$(1-\beta(t))\frac{\lambda}{\mu}(1-\alpha(t))$	$\beta(t)$	$(1-\frac{\lambda}{\mu})(1-\beta(t)))$
$x = \frac{B}{\_}$	$(1-\gamma(t))\frac{\lambda}{\mu}\alpha(t)$	$(1-\gamma(t))\frac{\lambda}{\mu}(1-\alpha(t))$	$\gamma(t)$	$(1-\frac{\lambda}{\mu})(1-\gamma(t)))$
$x = \frac{\_}{B}$	$(1-\beta(t))\frac{\lambda}{\mu}\alpha(t)$	$(1-\beta(t))\frac{\lambda}{\mu}(1-\alpha(t))$	$\beta(t)$	$(1-\frac{\lambda}{\mu})(1-\beta(t))$

Table 4.1: The transition probabilities $P_{\lambda,\mu}(x \to y)$ in the TKF1 model with $\alpha(t) := e^{-\mu t}$, $\beta(t) := \frac{\lambda(1-e^{(\lambda-\mu)t})}{\mu - \lambda e^{(\lambda-\mu)t}}$ and $\gamma(t) := 1 - \frac{\beta(t)\mu}{\lambda(1-e^{-\mu t})}$.

The TKF1 model fits into the concept of a *pair hidden Markov model* (*pair-HMM*) as described in Durbin *et al.* (1998). The usual HMM is a Markov chain which is hidden from an observer in the sense that he can only observe a sequence of "emissions" which depend on the current states. In the *pair*-HMM setting the observer sees a pair of sequences instead of a single sequence. In our case, the Markov chain is the bare alignment (corresponding to the path in the graphical representation) and the emissions are the DNA sequences. Durbin *et al.* (1998) describe various pair-HMMs that can be used for sequence alignment. In general, these models are constructed rather heuristically and are not exactly compatible with any stochastic sequence evolution model, as it is the case for the pair-HMM that arises from the TKF1 model.

Many algorithms for HMMs can be adapted to pair-HMMs. One of them is the *forward algorithm* for the calculation of the likelihood of possible values for the transition probabilities of the Markov chain, given the emitted sequence. In our case, this likelihood is the probability $p_{s,\lambda,\mu}(s_1, s_2)$ that the observed sequences s_1 and s_2 are emitted from a TKF1 pair-HMM with parameters s, λ, and μ (here we think of the time span t for the evolution of one sequence into the other to be scaled to unit time.)

For the computation of this likelihood it is necessary to sum over all possible alignments of the sequences:

$$p_{s,\lambda,\mu}(s_1, s_2) = \sum_{\text{bare alignment } w} p_{\lambda,\mu}(w) \cdot p_s(s_1, s_2 \mid w) \qquad (4.1)$$

$p_{\lambda,\mu}(w)$ is the probability of the bare alignment w for given insertion and deletion rates λ and μ, and $p_s(s_1, s_2 \mid w)$ is the probability of the sequences, given the bare alignment w and the substitution rate s.

The forward algorithm uses a dynamic programming approach to calculate the above sum. The main idea is the following (for details see Thorne *et al.* (1991) and Durbin *et al.* (1998)): Let the evolution parameters s, λ, and μ and a pair of sequences s_1 and s_2 of lengths n and m be fixed. For $0 \leq i \leq n$ and $0 \leq j \leq m$ let $f(i, j, {}^{\text{B}}_{\text{B}})$ be the probability that the following three events occur in a run of a TKF1 pair-HMM with the given parameters.

- The first i bases of the emitted sequence 1 coincide with the first i bases of s_1.

60

- The first j bases of the emitted sequence 2 coincide with the first j bases of s_2.

- In the bare alignment given by the hidden state sequence the i-th site of the first sequence is homologous to the j-th site of the second sequence.

$f(i, j, \substack{B \\ \_})$ is defined similarly, except that site i is not homologous to site j but is aligned with a gap between sites j and $j+1$ of sequence 2. Note that because of the Markov property of the bare alignment, one can easily compute $f(i, j, \cdot)$ if the values for $f(i-1, j-1, \cdot)$, $f(i-1, j, \cdot)$, and $f(i, j-1, \cdot)$ are known. For example, if there is a C at site i in the first given sequence and a G at site j in the second sequence (for $i, j > 1$), it is easy to see that the following equations hold:

$$
\begin{aligned}
f(i, j, \substack{B \\ \_}) = \Bigg[\; &f(i-1, j, \substack{B \\ B}) \cdot P(\substack{B \\ B} \to \substack{B \\ \_}) \\
&+ f(i-1, j, \substack{B \\ \_}) \cdot P(\substack{B \\ \_} \to \substack{B \\ \_}) \\
&+ f(i-1, j, \substack{\_ \\ B}) \cdot P(\substack{\_ \\ B} \to \substack{B \\ \_}) \Bigg] \cdot \pi_C
\end{aligned}
\tag{4.2}
$$

$$
\begin{aligned}
f(i, j, \substack{B \\ B}) = \Bigg[\; &f(i-1, j-1, \substack{B \\ B}) \cdot P(\substack{B \\ B} \to \substack{B \\ B}) \\
&+ f(i-1, j-1, \substack{B \\ \_}) \cdot P(\substack{B \\ \_} \to \substack{B \\ B}) \\
&+ f(i-1, j-1, \substack{\_ \\ B}) \cdot P(\substack{\_ \\ B} \to \substack{B \\ B}) \Bigg] \cdot \pi_C \cdot (1 - e^{-st}) \cdot \pi_G
\end{aligned}
\tag{4.3}
$$

Therefore the $n \cdot m \cdot 3$ values of the function $f(\cdot, \cdot, \cdot)$ can be computed efficiently by calculating $f(i, j, \cdot)$ iteratively while increasing i and j from $i = j = 0$ up to $(i, j) = (n, m)$. The desired likelihood is then:

$$
\begin{aligned}
p_{s, \lambda, \mu}(s_1, s_2) = \; &f(n, m, \substack{B \\ B}) \cdot P(\substack{B \\ B} \to \text{End}) \\
&+ f(n, m, \substack{B \\ \_}) \cdot P(\substack{B \\ \_} \to \text{End}) \\
&+ f(n, m, \substack{\_ \\ B}) \cdot P(\substack{\_ \\ B} \to \text{End})
\end{aligned}
\tag{4.4}
$$

61

Thorne *et al.* (1991) suggest using classical optimization algorithms in combination with the forward algorithm to search for the triple (s, λ, μ) that maximizes $p_{s,\lambda,\mu}(s_1, s_2)$ for given s_1 and s_2. Recently, Hein *et al.* (2000) have improved the efficiency of the algorithm. Note that the maximum likelihood estimator, which is obtained by this procedure, does not rely on a single alignment: it takes every possible alignment of the given sequences into account.

4.3 The sampling of pairwise alignments

The method

The TKF1 model can now be used to address the questions about the distribution of the model parameters and the reliability of alignment regions that were raised in the introduction. Let θ be the parameters of the model, i. e. in our case $\theta = (s, \lambda, \mu)$. Again, we take $t = 1$ for simplicity. We assume a Bayesian framework where we put some prior probability $\pi(d\theta)$ on the parameters. For given θ, we denote the probability of an alignment A by $p_\theta(A)$. Thus, we can attack those questions through *joint* sampling of the alignments A and the model parameters θ with weights proportional to $p_\theta(A)\pi(d\theta)$, where A runs through the alignments compatible with s_1, s_2. Using the idea of Gibbs sampling, we achieve this by a Markov chain Monte Carlo (MCMC) method in which alternately the alignment is sampled given the model parameters, and vice versa. As a prior for θ we propose $\pi(d\theta) = e^{-s}ds\, e^{-\lambda}d\lambda\, e^{-\mu}d\mu$; this makes the probability that no substitution occurs at a given surviving site uniformly distributed on $[0, 1]$. We write $p(d\theta, A)$ for $p_\theta(A)\pi(d\theta)$. Our task is to sample (θ, A) according to $p(d\theta, A|\, s_1, s_2)$.

If θ is fixed and we want to sample an alignment according to $p_\theta(A\,|\,s_1, s_2)$, then we can apply a classical HMM backward sampling algorithm (cf. Durbin *et al.*, 1998, pp. 89-99), using the function f defined above. With f we can easily compute the probability distribution for the last alignment state before **End** given the sequences and the mutation parameters. For $x \in \{ \begin{smallmatrix} B \\ B \end{smallmatrix}, \begin{smallmatrix} B \\ \_ \end{smallmatrix}, \begin{smallmatrix} \\ \_ \\ B \end{smallmatrix} \}$ the probability that the last state is x, is $c \cdot f(n, m, x) \cdot P(x \rightarrow \textbf{End})$, where c is a normalizing constant such that the three values for the tree states add up to 1. Therefore we can easily draw the last state at random according to this probability distribution. Assume for example we drew a $\begin{smallmatrix} B \\ \_ \end{smallmatrix}$ for the last state. In the graphical representation of the alignment, this would mean that the last edge is a vertical one from $(n-1, m)$ to (n, m). For all states $y \in \{ \begin{smallmatrix} B \\ B \end{smallmatrix}, \begin{smallmatrix} B \\ \_ \end{smallmatrix}, \begin{smallmatrix} \\ \_ \\ B \end{smallmatrix} \}$, given that the last state is $\begin{smallmatrix} B \\ \_ \end{smallmatrix}$, the probability that the preceding state equals y is $c' \cdot f(n-1, m, y) \cdot P(y \rightarrow \begin{smallmatrix} B \\ \_ \end{smallmatrix})$, where c' is a normalizing constant again. We draw the preceding state according to this distribution and continue in the same manner until we end up at the first state, i.e. until we have drawn an edge in the alignment graph that starts at vertex $(0,0)$.

On the other hand, if an alignment is given, we can use a Metropolis-Hastings approach (cf. Gamerman, 1997) for sampling the parameter θ according to $p(d\theta| A)$. Starting with some parameter vector θ_0 we generate a Markov chain on the parameter space in the following manner: If the current state in step i is $\theta_i = (s_i, \lambda_i, \mu_i)$, then generate independent random numbers $\widetilde{s}$, $\widetilde{\lambda}$, and $\widetilde{\mu}$, exponentially distributed with means s_i, λ_i, and μ_i. Consider $\widetilde{\theta} = (\widetilde{s}, \widetilde{\lambda}, \widetilde{\mu})$ as a proposal for the next state $(s_{i+1}, \lambda_{i+1}, \mu_{i+1})$. Then make a random decision: Accept it with probability

$$\min \left\{ 1 \, , \, \frac{p(d\widetilde{\theta}| A) \cdot s_i \cdot \mu_i \cdot \lambda_i}{p(d\theta_i| A) \cdot \widetilde{s} \cdot \widetilde{\mu} \cdot \widetilde{\lambda}} \cdot \exp\left(\frac{\widetilde{s}}{s_i} - \frac{s_i}{\widetilde{s}} + \frac{\widetilde{\lambda}}{\lambda_i} - \frac{\lambda_i}{\widetilde{\lambda}} + \frac{\widetilde{\mu}}{\mu_i} - \frac{\mu_i}{\widetilde{\mu}} \right) \right\} \quad (4.5)$$

or set $\theta_{i+1} = \theta_i$ if you do not accept[2]. As can be checked in a straightforward way, the posterior distribution for (s, λ, μ) is a (reversible) equilibrium for

---

[2]In the implementation we use a slightly different proposal chain, where the quotient

the process described above. By irreducibility, the process converges to this equilibrium distribution and can thus be used (at least approximatively) for sampling mutation parameters for given alignments. Of course, if one samples from one realization of the process, the results are not independent. However, the dependencies become small if one chooses the intervals between two samplings sufficiently large. Especially before the first sampling one should allow enough steps for the so called "burn-in", because the initial state could have been a bad guess in a region of low probability.

Now that we have an alignment sampling strategy for given mutation parameters and a mutation parameter sampling strategy for given alignments, we can combine them using the idea of Gibbs sampling (cf. Gamerman, 1997) and obtain a method for sampling mutation parameters and alignments simultaneously. We construct a Markov chain on the space of mutation parameters and alignments as follows: If (θ, A) is the current state, then sample a new alignment A' for the parameter triple θ, and afterwards perform a Metropolis-Hastings run to obtain a new parameter triple θ' given the alignment A'.

The disadvantage of this method is that we always have to sample alignments for new mutation parameters. For doing so, we have to compute the $3 \cdot n \cdot m$ values for the function f with the new parameters each time, which is very time consuming in general. Therefore, instead of sampling the whole alignment in each step, we realign only a part of it, about 30 nucleotides – then we have to compute only about 2700 values for f in each step. We use a part of the saved runtime to increase the number of steps between the samplings to compensate the dependencies in the alignments that arise from this strategy. Note that the posterior probability distribution $p(d\theta, A \mid s_1, s_2)$

$\lambda/(\mu - \lambda)$ is exponentially distributed (with the old value as expectation) instead of μ.

is still a reversible equilibrium for the Markov chain on $\{(\theta, A)\}$ which we obtain by the algorithm described above. Therefore we can use this process for simultaneous MCMC sampling of alignments and mutation parameters from their common posterior distribution.

As in most applications of MCMC methods, appropriate algorithm parameters as the number of steps in the sampling intervals, the duration of burn-in, and the length of alignment resampling ranges can only be found with experience gained from careful analysis of the sampling results.

An example: pseudogene/gene pairs

Pseudogene	Acc.	Pseudogene	Acc.	Pseudogene	Acc.	
1. β actin $\psi 1$	V00479	24. Cytochrome c ψC	M22880	49. Casein kinase II-α ψ	X64692	
2. β actin $\psi 2$	V00481	25. Cytochrome c ψE	M22886	50. Keratin 19 ψ	M33101	
3. β actin ψ	M55014	26. Cytochrome c ψF	M22889	53. HSC-70 ψ	Y00481	
4. γ actin ψ	M55082	27. Cytochrome c ψH	M22891	55. Ferredoxin ψ-A	M34787	
6. γ actin $\psi 1$	X04224	28. Cytochrome c ψI	M22892	56. Ferredoxin ψ-B	M34789	
7. Aldolase reductase ψ	M84454	29. α enolase ψ	X15277	57. Ferritin H ψ	J04755	
8. Cyclophilin $\psi 133$	X52856	30. ARS $\psi 3$	K01846	58. Tubulin-β $\psi 67$	M38484	E
9. Cyclophilin $\psi 167$	X52858	31. ARS $\psi 1$	K01845	60. Tubulin-β $\psi 14$P	K00840	D
10. Cyclophilin $\psi 18$	X52855	32. Aldolase B ψ	M21191	62. Tubulin-β $\psi 21$P	K00841	B
11. Cyclophilin $\psi 192$	X52857	33. Na/K-ATPase-β ψ	M25159	63. Tubulin-β $\psi 46$P	J00317	C
12. Cyclophilin $\psi 29$	X52853	38. D2-type cyclin ψ	M91003	64. Tubulin-β $\psi 7$P	K00842	F
13. Cyclophilin $\psi 39$	X52852	40. LDH-A ψ	X02153	66. TPI $\psi 5$A	K03224	
14. Cyclophilin $\psi 43$	X52854	41. LDH-B ψ	M60601	67. TPI $\psi 19$A	K03225	
15. Cytochrome c $\psi 1$	D00266	42. Lipocortin 2 ψA	M62895	68. TPI $\psi 13$C	K03223	
16. Cytochrome c $\psi 2$	D00267	43. Lipocortin 2 ψB	M62896	70. Prothymosin-α ψD	J04800	
17. Cytochrome c $\psi 3$	D00268	44. Lipocortin 2 ψC	M62898	71. Prothymosin-α ψF	J04801	A
18. Cytochrome c ψA	M22878	45. Metallothionein I ψ	M13073	72. Prothymosin-α ψG	J04802	
20. Cytochrome c ψG	M22890	46. Metallothionein II ψ	M13074	78. Adenylate kinase 3 ψ	X60674	
21. Cytochrome c ψJ	M22900	47. PGK ψX	K03201			
22. Cytochrome c ψK	M22893	48. PGK ψA	K03019			

Table 4.2: The analyzed pseudogenes. The marks refer to the labels in figure 4.1. The numbering is the same as in table 1 of Gu and Li (1995).

In order to compare the substitution and insertion rates obtained with our method with estimates of substitution rates and gap frequencies based on usual sequence alignments we applied it to the pseudogene/gene pairs analyzed by Gu and Li (1995). Pseudogenes seem to be appropriate for testing our approach as their evolution is not constrained by functional necessity. Since they stem from functional sequences they should not contain repeats of short sequence motifs which could distort the insertion-deletion-process locally. Their functional homologs on the other hand should be practically devoid of insertions and deletions. Although the main focus of Gu and Li (1995) is on the distribution of gap lengths, they also provide estimates of the number of differences per position. Lacking access to the alignments on which Gu and Li based their study, we retrieved the unaligned sequences and aligned the coding regions of the functional genes to the presumably homologous part of the corresponding pseudogene. For 20 sequence pairs the numbers and lengths of gaps given by Gu and Li were not compatible with the lengths of the available sequences; these were not analyzed further. The remaining 58 of the 78 pseudogenes treated are given in table 4.2. For each sequence pair the sampling procedure was started using 4/3 times the number of differences per position from Gu and Li (1995) as estimator of the substitution rate[3]. As an initial simple estimator of the insertion rate we used the number of gaps divided by twice the alignment length: $\frac{g}{L_x+L_y+g}$, where L_x and L_y are the sequence lengths and g is the number of gaps in Gu and Li (1995). With these values a first alignment was sampled. Then, after 10,000 burn-in runs, every thousandth of 100,000 runs was sampled. The range of resampling of the alignment in each of the steps was 30 nucleotides. Figure 4.1 compares the most probable sampled substitution rates with Gu and Li's values. In most cases the estimates given by Gu and Li lie well within the central 95% of the sampled values. This is even true for large substitution

---

[3]The factor 4/3 takes into account that a nucleotide can be replaced by a nucleotide of the same type in the TKF1 model.

rates. Most of our most probable substitution rates are bigger than Gu and Li's estimates. This may indicate that the score-optimization of alignments tends to make the aligned sequences more similar than they actually are (see chapter 2).

For the sequence pairs which are labeled A-F in table 4.2 the 97.5% quantiles of the sampled substitution rates are by far smaller than the estimates given by Gu and Li. The gap frequencies as well as the sizes of the gaps in the respective sampled alignments, however, were always close to the values given by Gu and Li (1995) (data not shown). Thus, any reasonable scoring system (including the human eye) would prefer our sampled alignments to Gu and Li's alignments. This calls for an explanation like a typing error in their paper or the use of a different sequence for the functional gene and cannot be due to an error in our method. Table 4.3 shows examples of sampled alignments. The small variability seen in the first two frames is typical. The erratic behavior in the last frame is probably due to a single long gap. This is of course not provided for in the assumptions of the TKF1 model.

4.4 Discussion

This chapter served mainly as an introduction into the fields of statistical alignments, hidden Markov models and Bayesian inference which are of importance for the understanding of the next chapter. We have seen how insertion-deletion models can be used to calculate the probability of a sequence alignment, to compute the joint probability of a pair of sequences or to get an idea of the reliability of alignment paths and of model parameters.

As far as the TKF1 model is concerned, one should note that many refinements are possible: Instead of the simple substitution model described above,

Paths	Most probable alignment
	CCAGTTGCGGAAGAAGAGGCA_CAGTCCAAAACAATAAGA CCAGTTGCGGAAGAAGAGGCAACAGTTCCAAACAATAAGA TCACTGTAGT TCACTGTAGT
	CGCCGATAGGATGCAGAAG___ATCACCACCCTGGCGCCC TGCCGACAGGATGCAGAAGGAGATCACTGCCCTGGCACCC AGCACAAT AGCACAAT
	GAGAAAGGCA___AGATTTTTGTTCCAAAGGGTGCCGCCC GAGAAAGGCAAGAAGATTTTTATTATGAAGTGT_TC_C_C AGTGCCACACCATGGAAAAGGGAA AGTGCCACACCGTTGAAAAGGGAG
	CTATATCCAGCAAGACACTAAGG__G__T_____GC_ TTATATCCAGCAAGACACTAAGGGCGACTACCAGAAAGCG _TG_T__ACC CTGCTGTACC

Table 4.3: Some typical samples of alignment paths together with the most probable of the sampled alignments. The paths and alignments shown are cutouts from the sampled alignments of LDH ψ, β actin ψ1, cytochrome c ψG and lipocortin 2 ψB.

one may plug in any of the reversible DNA or amino acid substitution models listed in the tables A.1 and A.2 and even allow for regional rate heterogeneity (Thorne and Churchill, 1995). All these could easily be incorporated into our sampling method, though this could lead to parameter over-fitting problems.

Our model's gravest defect is unfortunately not so easily remedied. The TKF1 model considers only insertions and deletions of single nucleotides. This assumption runs counter to a growing body of practical experience and it can lead to implausible alignments, as the last example in table 4.3 shows (see also Saitou and Ueda, 1994). The challenge is to find biologically more plausible models for the insertion-deletion process which still preserve the hidden Markov structure essential for computational feasibility. Thorne, Kishino and Felsenstein (1992) suggest a generalization of the TKF1 model (see chapter 5) that allows insertions and deletions of longer fragments, but they are forced to require that inserted fragments can only be deleted as a whole.

The programs that were applied in section 4.3 are freely available from: http://www.math.uni-frankfurt.de/~stoch/software/mcmcalgn.

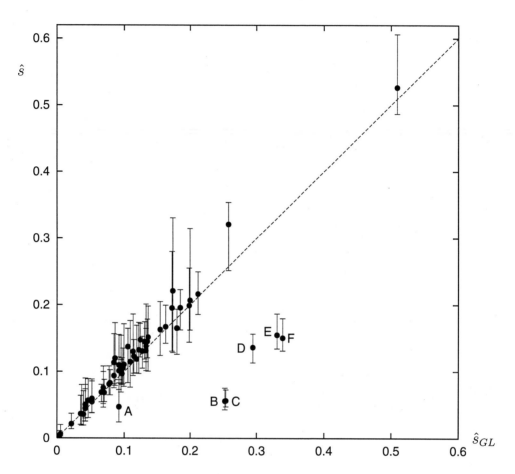

Figure 4.1: The most probable of the sampled substitution rates for each pseu-dogene/gene pair ($\hat{s}$) plotted against the values estimated by Gu and Li ($\hat{s}_{GL}$). The latter was evaluated as $-\ln\left(1 - \frac{4}{3}d\right)$ where d is the number of differences per position given in Gu and Li (1995). The dotted line is the identity, the plotted intervals indicate the 2.5% and the 97.5% quantiles. See table 4.2 for the meaning of the labels A-F.

Chapter 5

Simultaneous statistical multiple alignment and phylogeny reconstruction

5.1 Introduction

Statistical alignment procedures stand on the sound base of explicit models for the insertion and deletion process. Therefore applying them in the field of tree reconstruction is a tempting idea. Recently, the TKF1 model for two sequences has indeed been generalised leading to an algorithm which gives the probability of n sequences which have evolved on a star shaped tree (Steel and Hein, 2001) or on arbitrary trees (Lunter *et al.*, 2003). There now also is a method that samples multiple alignments according to their posterior probability under this extension of the TKF1 model conditioned on a given phylogeny and parameter values (Holmes and Bruno, 2001).

The TKF1 model, however, is not very realistic in that it only allows for insertions and deletions of single nucleotides (or amino acids). This deficiency was remedied in Thorne *et al.* (1992) by the TKF2 model. Here, we will first give a brief overview over the TKF2 model and provide a possible extension to the case of n sequences which are related by a tree (section 5.2). Then, we will utilize this model to simultaneously construct a multiple alignment, reconstruct the sequences' phylogeny and estimate the mutation parameters (section 5.3).

5.2   Model

If we want to describe the evolution dynamics which takes a DNA or protein sequence into another one we have to model the substitution process as well as the process of insertions and deletions. The substitution process we think of is simply any model where the positions evolve independently according to the same probabilistic dynamics and which has a unique stationary distribution. These are described in detail elsewhere (Tavaré, 1986; Strimmer, 1997; Müller and Vingron, 2000).

Here we focus only on the insertion and deletion process which is the aforementioned TKF2 model. Under this model every sequence is thought to be composed of disjoint fragments the lengths of which are geometrically distributed with expectation $1/(1 - \rho)$. Between any two fragments and at the ends of the sequence a new fragment which is drawn from the same distribution may be inserted with rate λ. At rate μ each fragment is deleted entirely. Thus, the fragmentation structure remains unchanged: the fragments are always deleted as a whole and are never split by an insertion. Although this may be considered unrealistic, the results in Metzler (2003) indicate that at least in the case of a sequence pair the estimation of the model parameters is quite robust against violation of this assumption. Obviously, the insertion-deletion process of the TKF2 model acts in the same way on the sequence fragments as does the TKF1 model on the nucleotides. Thus, we get an equilibrium for the number of possible insertion sites, i. e. the number of fragments plus 1, which is the geometric distribution with expectation $\mu/(\mu - \lambda)$ and thus we also have an equilibrium distribution for the sequence length. It is further assumed that the length of the ancestral sequence was taken from the equilibrium distribution of the process. In Thorne *et al.* (1991) and Thorne *et al.* (1992) it is also specified how a given realization of the indel process is to be represented as an alignment. The rule is to consider each inserted fragment as offspring of its left neighbor and to write it as close as possible to its ancestor in the alignment. The alignment notation rule and the fixed fragmentation structure are necessary to make a TKF2-generated sequence pair alignment, when read from left to right, a Markov chain on the states $\frac{B}{B}$, $\frac{B}{-}$, $\frac{-}{B}$, Start and End; see figure 5.1 and (Thorne *et al.*, 1991, 1992; Metzler, 2003, see also chapter 4) for details. This property is crucial for computational tractability since it allows to apply pair HMM procedures (Durbin *et al.*, 1998; Holmes and Bruno, 2001). Using these techniques we can compute the joint probability of two related sequences by summing over all paths through the HMM which might have produced them and we can compute the probability of any alignment of these sequences by walking through the states it contains.

If we want to extend the TKF2 model to n sequences which are related by a binary tree, there are two topics we have to consider: Firstly, think of a sequence X which is the most recent common ancestor of the sequences Y and Z (see the diagram in in the upper right corner of figure 5.2). If a fragment in X has children in Y as well as in Z, i. e. there have been insertions on both branches, the TKF2 model does not provide a rule in which order we should write down these independent insertions. Therefore, we have to add the convention that insertions on the left subtrees (the root being at the bottom of the tree) are recorded before the ones on the right subtrees in a depth-first tree traversal. The second problem we have to deal with is the fragmentation of the sequences. The HMMs in the figures 5.2, 5.3 and 5.4 keep the fragmentation structure fixed over the whole subtree under consideration. Although this is the somehow canonical extension of the TKF2 model it is probably unrealistic and also impractical. In order to compute the probability of a multiple alignment we would have to construct an HMM for the entire phylogeny whereas the HMM from figure 5.1 suffices if we confine the fragmentation structure to the individual branches. Moreover, the algorithms we are going to introduce in section 5.3 would require a change in this fixed fragmentation from time to time, yet it is not obvious how this can be done without getting into trouble. That is why we allow the fragmentation structure to change at every node of the tree. This is also a step towards a fragment insertion-deletion model at arbitrary sites. Nevertheless, as will be explained in section 5.3 we still use the multiple HMMs in the figures 5.2-5.4, which leave the fragmetation fixed over the nodes, and correct their output in subsequent steps.

5.3   Algorithm

Holmes and Bruno (2001) introduced a method that samples multiple alignments according to their posterior probability distribution under the TKF1 model. We will now present three search strategies that are intended to find the tree (topology and branch lengths), the multiple alignment and the model parameters that maximise this probability under the extended TKF2 model described above. Each of them is based on the simulated annealing scheme

74

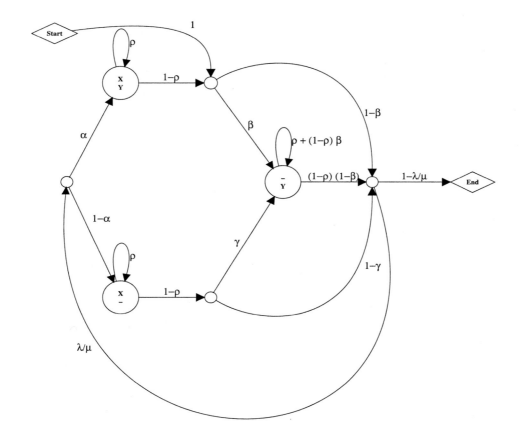

Figure 5.1: The TKF2 pair-HMM that produces two sequences, X and Y (X being the ancestor of Y). $\begin{smallmatrix}X\\Y\end{smallmatrix}$ is the 'match' state: The bases in the sequences are homologous. $\begin{smallmatrix}X\\-\end{smallmatrix}$ is the 'deletion' state: There is a base in sequence X which got deleted, thus there is a gap in sequence Y. $\begin{smallmatrix}-\\Y\end{smallmatrix}$ is the 'insertion' state: There is a base in sequence Y and a gap in sequence X. ρ is the parameter of the geometric distribution of fragment lengths, α is the probability that a fragment survives, β is the probability of more insertions given one or more extant descendants and γ is the probability of insertions given that a fragment did not survive. α, β and γ are functions of the time. See Holmes and Bruno (2001) for details on computing these values. Note that there are two ways from the 'match' state back to itself (the same is true for the 'deletion' state).

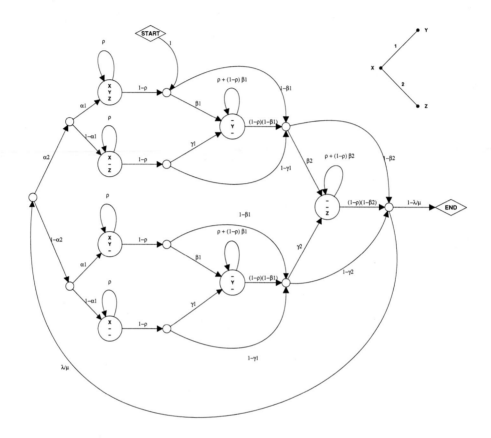

Figure 5.2: The fixed-fragmentation TKF2 evolutionary HMM that produces two sequences , Y and Z, and their most recent common ancestor, X (see the graph in the upper right corner). As the probabilities of fragment survival and of a fragment having offspring with or without itself surviving (α, β and γ resp.) depend on the time between an ancestral sequence and its descendant, they are indexed by the appropriate edge labels.

sketched in section 5.3.1. They only differ in the way they start and end.

A realisation of the extended TKF2 model for some phylogeny of n sequences produces sequences at the inner nodes and pairwise alignments of every sequence to its neighbors in the tree. When computing the probability of a multiple alignment of the sequences at the leaves of the tree we have to sum over all possible sequences at the inner nodes and all possible pairwise alignments along the tree's branches that yield this multiple alignment. Whereas summing over all possible base or amino acid assignments at the inner nodes of the tree can easily be done (Felsenstein, 1981), summing over all possible pairwise alignments is prohibitively slow and computation time increases drastically with the number of sequences (Steel and Hein, 2001). We therefore just like Holmes and Bruno (2001) resort to emitting sequences at the inner nodes of the tree when producing or judging (i. e. computing their probability) multiple alignments.

5.3.1   Strategy I: pure simulated annealing

This strategy starts with a random unrooted guide tree (one randomly chosen leaf being considered to be the root) and some arbitrary parameter values for the substitution and insertion-deletion processes according to which the sequences are aligned progressively. Then, two steps are repeated again and again until one does not observe significant changes in the posterior probability of the multiple alignment: In step 1 we propose a new tree and alignment, in step 2 we propose new parameters. The probability to accept bad proposals is decreased monotonically from iteration to iteration.

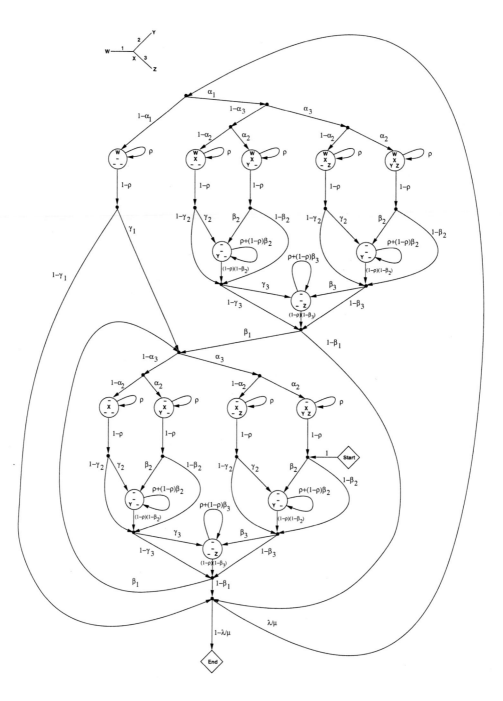

Figure 5.3: The fixed-fragmentation TKF2 evolutionary HMM for three sequences that are related by the tree depicted in the upper left corner of the figure. Sequence W is considered to be the ancestor.

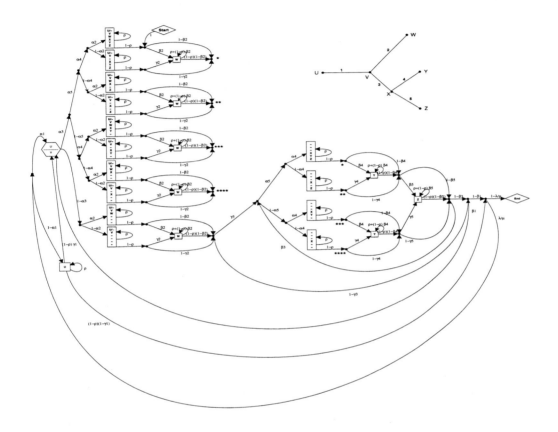

Figure 5.4: The fixed-fragmentation TKF2 evolutionary HMM for four sequences related by a binary tree (upper right corner). Sequence U is considered to be the ancestral sequence. Asterisks mark states which are identical (i. e. which should be connected by arrows pointing from left to right with the label '1'). The hexagon on the left is a "different input equal output" switch for states which either do or do not emit a base at the ancestral sequence.

Getting the initial alignment

The initial alignment is created by a progressive alignment from the tips of the randomly chosen guide tree downwards: A pair of sibling sequences is aligned (i. e. the most probable path for the given parameters is chosen). Then, given this alignment a sequence at their parent node is produced (together with its pairwise alignments to its children) by the HMM in figure 5.2. Then, this sequence is aligned to its sibling and so on. At the end of this procedure we have a multiple alignment of all sequences in the data set which consists of pairwise alignments along the branches of the tree.

Changing tree and alignment

Proposing a new tree is done in a similar way as in the tree sampling procedure outlined in Mau and Newton (1997): For every leaf i of the tree with branch length l_i a new branch length l_i^* is drawn from the uniform distribution $U(l_i - \delta, l_i + \delta)$ if the positive parameter δ is smaller than l_i and from $\frac{\delta - l_i}{\delta} U(0, \delta - l_i) + \frac{l_i}{\delta} U(\delta - l_i, \delta + l_i)$ if δ is bigger than l_i. Thus, it is guaranteed that a leaf never gets a negative branch length. The branch lengths of internal branches on the other hand are always drawn from $U(l_i - \delta, l_i + \delta)$ and may therefore become negative. In this case the branch is removed and one of the two alternative topologies is chosen randomly (the corresponding new branch gets the absolute value of the drawn branch length as length). Every time we change an internal branch we have to emit new sequences at the new inner nodes. This is done using the HMM from figure 5.4. We keep the alignment of the four 'neighbors' of this branch fixed and emit new sequences at the inner nodes (V and X in figure 5.4). The alignment is further refined by producing a new sequence at every internal node of the tree (in randomized order) using the HMM in figure 5.3 and by optimizing every pairwise alignment. The acceptance of the proposed tree depends on the likelihood ratio of the new and the old tree. In order not to propose only trees which will be rejected, the parameter δ is adjusted in the course of the program in a way that the acceptance rate is maximised. When computing the likelihood of such a multiple alignment we sum over all possible assignments of bases, yet we do not attempt to sum over all possible states of the TKF2 model.

When proposing the new alignment we use several multiple HMMs which keep the fragmentation structure fixed over different subtrees. However, this does not cause major complications, because when we decide between two tree topologies we only take their likelihoods into account, the proposal distribution does not matter.

Getting new parameters

New model parameters are suggested and accepted just in the same way as in chapter 4: Let $\theta = (s_1, s_2, ..., s_l, \lambda, \rho)$ be the model parameters, with s_1 to s_l being the various parameters of the substitution model, λ being the insertion rate and ρ being the parameter of the distribution of fragment lengths. As prior for θ we take $\prod_{\nu=1}^{l} e^{-s_\nu} ds_\nu \ e^{-\lambda} d\lambda \ e^{1-1/\rho} d\rho$. If the current parameter vector in step i is θ_i then we propose new parameters $\tilde{\theta}$ by drawing every component of $\tilde{\theta}$ except of $\tilde{\rho}$ from an exponential distribution whose mean is the corresponding component of θ_i. For the parameter of the fragment length we draw $1/\tilde{\rho} - 1$ from an exponential distribution whose mean is $1/\rho_i - 1$. Note that we do not attempt to optimize the insertion and the deletion rate independently. Instead we use a fixed ratio of the insertion and the deletion rates which yields the average sequence length as the expectation.

Simulated Annealing

We adopt the cooling schedule from Salter and Pearl (2001), i. e. a bad proposal x_j is accepted in step i with $\exp(\frac{-(f(x_i)-f(x_j))}{c_i})$ where the control parameter $c_i = \frac{U}{1+(i-1)b}$. Here U is the maximum jump of f in one step and is determined in an initial 'burn-in' phase. The parameter b equals $\frac{c}{(1-a)n-a\frac{\ln L}{m}}$ where n is the number of sequences, m the number of sites in the alignment at the end of the 'burn-in', $\ln L$ this alignment's loglikelihood and c and a are values between 0 and 1 (for the simulation and the example presented in section 5.4 both were set to 0.5). The tree optimization and the parameter optimization have both a cooling scheme of their own. Thus, x_i and x_j stand for the old and the proposed parameter values respectively or for the old and the proposed tree. In the former case $f(x_i)$ has to be interpreted as the natural logarithm of the product of the probability of the old parameters given the alignment and the probability to propose x_j when in x_i. In the latter case $f(x_i)$ is the loglikelihood of tree x_i given the alignment.

5.3.2 Strategy II: simulated annealing with improved start

The pure simulated annealing strategy described in the last section might start in a very exotic area of tree space. Strategy II tries to start with a more reasonable tree which is found via neighbor-joining (Saitou and Nei, 1987). Therefore, a distance matrix for the unaligned sequences has to be estimated. This is done by a coarse variant of the EM-algorithm (for longer sequences we use Viterbi training, i. e. we just optimise the probability of the optimal path (Durbin *et al.*, 1998)) in Thorne and Churchill (1995): Alignment paths for each pair of sequences are sampled like in chapter 4, the transitions from one HMM state to the next and the numbers of the found matches and mismatches are counted. Then, the likelihood of the substitution rate and the weight of the indel process as compared to the substitution process given these observations is optimised. Thus, we have for each pair an estimate of the substitution rate and a ratio of the speeds of the indel and the substitution processes. The median of these "indel-substitution-ratios" is used to rescale the rates of the indel process and the whole estimation procedure is repeated, but now only to estimate the substitution rates (Thorne and Kishino, 1992). With these rates we then build the initial neighbor-joining tree (note that we do not try to estimate the parameters of the substitution process in this step).

5.3.3 Strategy III: simulated annealing with improved start and nearest-neighbor interchanges

As always with simulated annealing the procedure might run for a very long time without any changes in the estimates. Yet, we will never be sure if there will not be any jumps some time in the future. To explore tree space a bit faster we do nearest-neighbor interchanges (Swofford *et al.*, 1996) in these phases of stagnation: Every internal branch is visited in a random order and pairwise distances for its four adjacent nodes are estimated (to remove the least influence of the old topology, new sequences have been emitted at these nodes by the HMM from figure 5.2 just paying regard to the subtree which starts from there). The distance estimation is done by the EM-algorithm described above (without rescaling of the insertion rate). Then, we reconstruct a neighbor-joining tree for these four nodes and emit new sequences at the new internal nodes. As this procedure does not pay respect to the new tree's likelihood and is only distance based the new tree might well have a drastically reduced likelihood. After having done the nearest-neighbor interchanges the program returns to the simulated annealing schedule until it encounters another phase of stagnation. The program quits after having passed through a prespecified number of nearest-neighbor interchange steps saving the most probable of the found alignments. Like strategy II this strategy also starts with a neighbor-joining tree.

5.4   Application

5.4.1   Simulation

A four-taxon example

In a first simulation we simulated 100 data sets of expected sequence length 1000 along the phylogeny shown in figure 3.3. The setting for this simulation were exactly the same as the ones in section 3.3 which means that we used the TKF1 model together with the Jukes-Cantor substitution model. Given the four unaligned sequences at the leaves of the tree we tried to reconstruct their phylogeny using strategy III together with the Felsenstein substitution model and an initial insertion rate of 0.1 per time unit. 100 iterations at the beginning of the search were the burn-in phase. If the absolute value of the difference of the likelihoods in two succesive iterations was smaller than 1 for more than 1000 iterations, we conducted nearest-neighbor interchanges after which we continued with the simulated annealing. We stopped the searches when this situation occurred for the 9th time. For 95 of the 100 data sets we were able to reconstruct the correct topology which is quite an improvement over the results from section 3.3.

84

An eight-taxon example

To study the behaviour of the three search strategies we simulated 100 data sets of nucleotide sequences on the tree shown in figure 3.5B (branch length a being 0.01 and branch length c equalling 0.07) with the expected length of 500 nucleotides under Kimura's two-parameter model (Kimura, 1980) with the transition transversion parameter being 4.0. The insertion rate of the TKF2 process was 0.002 and the expected fragment length was 10 ($\rho = 0.9$). As described in section 4.2 the fragmentation structure was allowed to vary from branch to branch.

Given the unaligned leaf sequences of the simulated data sets we then tried to estimate the underlying phylogeny, their alignments and the model parameters with each of three methods from section 5.3 using the Tamura-Nei substitution model (Tamura and Nei, 1993) (for equal base frequencies and with a pyrimidine-purine-transition parameter of 1 this model is equivalent to Kimura's two-parameter model). The initial value for the transition-transversion parameter was 2.0 and the pyrimidine-purine-transition parameter initially was 1.0. The initial insertion rate was 0.1 and the initial expected fragment length was 2 ($\rho = 0.5$). 1000 iterations at the beginning of the search were the burn-in phase (see 5.3.1). If the absolute value of the difference of the likelihoods in two succesive iterations was smaller than 5 for more than 800 iterations strategy III conducted nearest-neighbor interchanges after which it continued with the simulated annealing. We stopped the searches when this situation occurred for the 17th time.

Table 5.2 shows that all three strategies performed equally well in finding the alignment. On average more than 96% of the positions in the true alignments were also present in the reconstructed ones. The parameter estimates (see tables 5.3 and 5.4) are also quite reasonable, no matter which of the three

Strategy	Mean partition distance from the tree in fig. 3.5B
I	7.8 ± 2.3
II	5.3 ± 2.2
III	2.0 ± 2.4

Table 5.1: The mean ± the standard deviation of the partition distances between the best trees found by the three search strategies and the true phylogeny of the simulated data sets.

Strategy	Mean proportion of correct alignment positions
I	0.962 ± 0.0238
II	0.973 ± 0.0156
III	0.977 ± 0.0142

Table 5.2: The mean ± the standard deviation of the proportion of alignment positions in the simulated data sets that were also found in the corresponding most probable alignments.

search strategies is used. In their ability to reconstruct the phylogeny (table 5.1) the three strategies show, however, remarkable differences with strategy III performing best. As a measure of the dissimilarity of the tree from figure 3.5B and the reconstructed trees we computed their partition distance (Robinson and Foulds, 1981). In order to interprete these results one should keep in mind that the expected partition distance of two randomly chosen bifurcating 8-taxon topologies is 9.4 and that less than 1% of those random topology pairs have a partition distance smaller than 6 (Penny *et al.*, 1982).

Strategy	Mean $\hat{\tau}$ ($\tau = 4.0$)	Mean $\hat{\kappa}$ ($\kappa = 1.0$)
I	4.07 ± 0.84	1.06 ± 0.21
II	3.98 ± 0.76	1.05 ± 0.23
III	3.96 ± 0.83	1.04 ± 0.22

Table 5.3: The mean of the estimates of the substitution model parameters ($\pm$ their standard deviations).

Strategy	Mean $\hat{\lambda}$ ($\lambda = 0.002$)	Mean $\hat{\rho}$ ($\rho = 0.9$)
I	$2.1 \times 10^{-3} \pm 1.1 \times 10^{-3}$	0.85 ± 0.056
II	$1.7 \times 10^{-3} \pm 8.1 \times 10^{-4}$	0.87 ± 0.044
III	$2.1 \times 10^{-3} \pm 1.2 \times 10^{-3}$	0.88 ± 0.041

Table 5.4: The mean of the estimates of the parameters of the TKF2 model ($\pm$ the standard deviation).

5.4.2   An example: HVR-1 from primates

In order to see whether our method works with real data we applied strategy III to a data set of HVR1 sequences from 12 primate species. Details on the accession numbers are given in the legend of figure 5.5. We modelled the substitution process with the Tamura-Nei model starting with the transition-transversion parameter being set to 5.0 and the pyrimidine-purine-transition parameter initially being 1.0. The initial expected fragment length was 2 ($\rho = 0.5$). 100 iterations at the beginning of the search were the burn-in phase (see 5.3.1). If the absolute value of the difference of the likelihoods in two succesive iterations was smaller than 5 for more than 1200 iterations we conducted nearest-neighbor interchanges after which we continued with the simulated annealing. We stopped our search when this situation occurred for the 25th time.

Figure 5.5 shows the consensus tree of the best trees found in 100 independent runs of strategy III. The edge labels show the percentage of trees which contain the respective branch. These values should not so much be seen as support values for the individual branches, but rather as an illustration of how good we were in finding the optimum. Although the depicted tree conforms well with textbook views on primate phylogeny, it has two striking features. Firstly, the gorilla sequence, which has at least one very long deletion (Burckhardt *et al.*, 1999), seems to be misplaced and should rather be the sister group to a Homo-Pan cluster. Secondly, the relationship among humans, chimps and the Pongo-Gorilla cluster remains unresolved. 31 of the reconstructed trees favour a Pongo-Gorilla-Homo cluster and 31 trees exhibit a Pongo-Gorilla-Pan cluster (data not shown). This may, however, be a side effect of the misplacement of the gorilla sequence. The estimates of the model parameters are given in table 5.5.

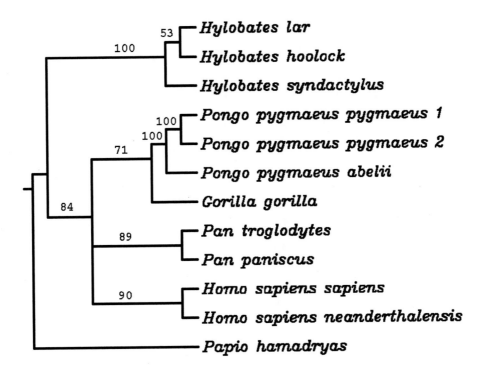

Figure 5.5: Consensus tree of the hominoid HVR1 phylogenies with the highest likelihood found in 100 independent runs of strategy III. The accesion numbers and included positions of the sequences are as follows: NC_002082 1-366 (*H. lar*), AF311725 1-378 (*H. hoolock*), AF311722 1-396 (*H. syndactylus*), Hvr1_ID_1 2-379 (*H. sapiens neanderthalensis*), Hvr1_ID_1244 24-400 (*H. sapiens sapiens*), Hvr1_ID_264 24-398 (*P. troglodytes*), Hvr1_ID_261 24-399 (*P. paniscus*), Hvr1_ID_389 1-374 (*P. pygmaeus pygmaeus 1*), Hvr1_ID_390 1-374 (*P. pygmaeus pygmaeus 2*), Hvr1_ID_388 1-372 (*P. pygmaeus abelii*), Hvr1_ID_262 24-326 (*G. gorilla*), NC_001992 1-378 (*P. hamadryas*). Accession numbers which begin with 'Hvr1_ID_' are the identifiers in the hvrbase (Handt *et al.*, 1998). Regions with high similarity were determined with the help of the program dotter (Sonnhammer and Durbin, 1996).

parameter	mean ± std.dev.
τ	1.94 ± 0.20
κ	1.03 ± 0.12
λ	$1.33 \times 10^{-3} \pm 5.4 \times 10^{-4}$
ρ	0.798 ± 0.043

Table 5.5: The estimated parameters for the HVR1 data set. τ is the Tamura-Nei model's transition-transversion parameter, κ the pyrimidine-purine-transition parameter, λ the insertion rate and ρ the parameter of the distribution of fragment lengths. The table shows the averages over 100 independent runs of strategy III

5.5 Discussion and Outlook

In this chapter we have outlined methods which allow the statistical inference of a multiple alignment, a phylogenetic tree with branch lengths and the parameters of the substitution and the insertion and deletion process. The algorithms can easily be adapted to more complex and more realistic insertion and deletion models and also to allow for rate heterogeneity (cf. Thorne and Churchill, 1995). A more pressing task is however decreasing the run time of our programs and to graze the parameter and tree spaces more efficiently. A step in this direction would already be to carefully analyze and optimize parameters like the number of burn-in and sampling steps in the simulated annealing procedure. Another step would be adapting some of the numerous tree search heuristics to this problem. We hope that further optimization will make it possible to develop methods to judge the reliability of multiple alignments and phylogeny reconstruction by Bayesian MCMC-sampling, at least if the number of sequences is not too high.

Chapter 6

Summary

The focus of this thesis was the influence that the optimization of sequence alignments may exert on the estimation of substitution rates and on the reconstruction of phylogenetic trees. A topic which up to now has not gained much attention.

Using the concept of parametric sequence alignment, we could show in chapter 2 that substitution rates cannot be estimated reliably from optimal sequence alignments. Therefore, we developed a simple least-squares method which lead to better distance estimates in a large area of the parameter space.

The simulations of chapter 3 exemplified the problems that may arise when one reconstructs trees from multiple alignments. We observed a drastic reduction of the efficiency of neighbor-joining even if the sequences were reasonably long. We also saw that multiple alignment programs may force their implicit assumptions upon the sequences under study, thus being the cause of erroneous trees and erroneous support values.

Finally, we studied the possibilities to make use of insertion-deletion models. In the last chapter, we introduced the first method which applies a fragment insertion-deletion model to simultaneously reconstruct a phylogeny and a multiple alignment for a set of sequences. The development of this method also resulted in writing a substantial software tool.

Appendix A

Program manuals

A.1 ALIFRITZ

The program ALIFRITZ is a C++ implementation of the algorithm described in chapter 5. Its source code is available on request. Given a guide tree and a set of sequences, the program searches for the combination of tree topology, branch lengths, multiple alignment and parameter values which has maximal posterior probability. It is run from the command line by typing 'alifritz control x' where 'control' stands for the name of a file which holds the parameter settings and where 'x' is either 0, 1 or 2 and selects one of the three following search strategies: simulated annealing starting with a random tree, simulated annealing starting with a neighbor-joining tree and simulated annealing starting with a neighbor-joining tree and with nearest-neighbour interchanges (see chapter 5). As far as the modelling of sequence evolution is concerned, the user may choose among two insertion-deletion models, seven nucleotide and five amino acid substitution models. The user can also specify which parameters are to be estimated by the program and which ones should remain unchanged. In order to document the parameter settings of a run of ALIFRITZ and in order to keep this record comprehensible, the various program options are controlled by a simple ANSI-text file that consists of a series of blocks which group the user-specified settings. Furthermore, this control file allows for comments which are started with a hash ('#'), end with a new line and may appear anywhere in the file. Blocks begin with the name of the block, which is followed by a colon and a new line. After this initialization, a block has several lines, each of which assigns a value to one of the block's variables. These assignments are done with the '=' operator. Depending on the variable, the assigned value may either be a number, a string or a vector of positive numbers which begins and ends with round brackets and whose entries are separated by commas. Every block and each of the statements it contains may be omitted as the program has default settings for every parameter. We tried to keep the conventions for the control file as flexible as possible. Thus, ALIFRITZ does not care about the case of letters or about whitespaces other than new lines.

The 'SUBST' block:

This block has three variables that specify the substitution process: 'model', 'parameter' and 'freqs'. By assigning one of several keywords to the variable 'model', the user can choose one of the substitution models listed in the tables A.1 and A.2. The initial rate parameters of the chosen model are set by assigning a vector to the variable 'parameter'. The input order and the default values of the parameters of the nucleotide substitution models are shown in table A.1. To the variable 'freqs', the user may assign a vector of nucleotide or amino acid frequencies. These frequencies need not sum up to one, as the program will do the normalization. Therefore, however, the vector has to have an entry for every nucleotide or amino acid respectively.

The order of the frequencies corresponds to the alphabetical order of the usual symbols for nucleotides or amino acids. If one wants to estimate the frequencies from the data one may either write 'freqs = estimate', 'freqs = default' or omit an assignment to 'freqs'.

Table A.1: The nucleotide substitution models as implemented by ALIFRITZ. See the end of the table for a detailed explanation.

model keyword, parameters (input order and default values)	rate matrix
Jukes-Cantor model (Jukes and Cantor, 1969) JC69	$\begin{array}{c}\\ A\\ C\\ G\\ T\end{array}\begin{array}{cccc}A & C & G & T\\ \left(\begin{array}{cccc}\cdot & \frac{v}{4} & \frac{v}{4} & \frac{v}{4}\\ \frac{v}{4} & \cdot & \frac{v}{4} & \frac{v}{4}\\ \frac{v}{4} & \frac{v}{4} & \cdot & \frac{v}{4}\\ \frac{v}{4} & \frac{v}{4} & \frac{v}{4} & \cdot\end{array}\right)\end{array}$
Felsenstein model (Felsenstein, 1981) FEL81	$\begin{array}{c}\\ A\\ C\\ G\\ T\end{array}\begin{array}{cccc}A & C & G & T\\ \left(\begin{array}{cccc}\cdot & v\pi_C & v\pi_G & v\pi_T\\ v\pi_A & \cdot & v\pi_G & v\pi_T\\ v\pi_A & v\pi_C & \cdot & v\pi_T\\ v\pi_A & v\pi_C & v\pi_G & \cdot\end{array}\right)\end{array}$
Kimura's 2-parameter model (Kimura, 1980) KIMURA2P $\tau = 1/2$	$\begin{array}{c}\\ A\\ C\\ G\\ T\end{array}\begin{array}{cccc}A & C & G & T\\ \left(\begin{array}{cccc}\cdot & \frac{v}{4} & \frac{v\tau}{2} & \frac{v}{4}\\ \frac{v}{4} & \cdot & \frac{v}{4} & \frac{v\tau}{2}\\ \frac{v\tau}{2} & \frac{v}{4} & \cdot & \frac{v}{4}\\ \frac{v}{4} & \frac{v\tau}{2} & \frac{v}{4} & \cdot\end{array}\right)\end{array}$

continued on next page

Kimura's 3-parameter model (Kimura, 1981) KIMURA3P $\tau = 1/2, \kappa = 1$	$\begin{array}{c} \\ A \\ C \\ G \\ T \end{array} \begin{array}{cccc} A & C & G & T \\ \left(\begin{array}{cccc} \cdot & \frac{v}{4} & \frac{v\tau}{\kappa+1} & \frac{v}{4} \\ \frac{v}{4} & \cdot & \frac{v}{4} & \frac{v\tau\kappa}{\kappa+1} \\ \frac{v\tau}{\kappa+1} & \frac{v}{4} & \cdot & \frac{v}{4} \\ \frac{v}{4} & \frac{v\tau\kappa}{\kappa+1} & \frac{v}{4} & \cdot \end{array} \right) \end{array}$
HKY model (Hasegawa *et al.*, 1985) HKY $\tau = 1/2$	$\begin{array}{c} \\ A \\ C \\ G \\ T \end{array} \begin{array}{cccc} A & C & G & T \\ \left(\begin{array}{cccc} \cdot & v\pi_C & 2v\tau\pi_G & v\pi_T \\ v\pi_A & \cdot & v\pi_G & 2v\tau\pi_T \\ 2v\tau\pi_A & v\pi_C & \cdot & v\pi_T \\ v\pi_A & 2v\tau\pi_C & v\pi_G & \cdot \end{array} \right) \end{array}$
Tamura-Nei model (Tamura and Nei, 1993) TAMNEI $\tau = 1/2, \kappa = 1$	$\begin{array}{c} \\ A \\ C \\ G \\ T \end{array} \begin{array}{cccc} A & C & G & T \\ \left(\begin{array}{cccc} \cdot & v\pi_C & \frac{4v\tau}{\kappa+1}\pi_G & v\pi_T \\ v\pi_A & \cdot & v\pi_G & \frac{4v\tau\kappa}{\kappa+1}\pi_T \\ \frac{4v\tau}{\kappa+1}\pi_A & v\pi_C & \cdot & v\pi_T \\ v\pi_A & \frac{4v\tau\kappa}{\kappa+1}\pi_C & v\pi_G & \cdot \end{array} \right) \end{array}$
general reversible model (Lanave *et al.*, 1984) GENREV $\alpha = 1, \beta = 1, \gamma = 1,$ $\delta = 1, \varepsilon = 1, \zeta = 1$	$\begin{array}{c} \\ A \\ C \\ G \\ T \end{array} \begin{array}{cccc} A & C & G & T \\ \left(\begin{array}{cccc} \cdot & v\alpha\pi_C & v\beta\pi_G & v\gamma\pi_T \\ v\alpha\pi_A & \cdot & v\delta\pi_G & v\varepsilon\pi_T \\ v\beta\pi_A & v\delta\pi_C & \cdot & v\zeta\pi_T \\ v\gamma\pi_A & v\varepsilon\pi_C & v\zeta\pi_G & \cdot \end{array} \right) \end{array}$

Table A.1: The nucleotide substitution models as implemented by ALIFRITZ. The first column holds the model's names as well as the appropriate keyword in the control file. In the case that the model has parameters which can be set by the user, these parameters' input order and their default values are also given. The models' rate matrices are shown in the second column. The parameter v is set by the program in order to keep the processes' overall substitution rates equal to 0.01 and cannot be specified by the user. As usual, π_N stands for the frequency of base N and the '$\cdot$' are placeholders for the negative of the sum of the other row entries.

model	keyword
Dayhoff model (Dayhoff, 1978)	DAYHOFF
general reversible model (Müller and Vingron, 2000)	MV2000
JTT model (Jones *et al.*, 1992)	JTT
mtREV model (Adachi and Hasegawa, 1996)	MTREV24
BLOSUM62 model (Henikoff and Henikoff, 1992)	BLOSUM62

Table A.2: The implemented amino acid substitution models and the respective keywords.

The 'INDEL' block:

The block 'INDEL' comprises the settings for the insertion-deletion model. It has the variables 'model' and 'parameter'. By writing 'model = TKF1' the user chooses the TKF1 model and 'model = TKF2' makes the program apply the TKF2 model (see chapters 4 and 5). To the variable 'parameter', the user can assign a vector that embodies the initial settings for the insertion rate, the deletion rate and, in the case of the TKF2 model, the parameter of the fragment length distribution.

The 'KEEP' block:

In this block, the user may select model parameters whose values the program is not allowed to change. Unlike other blocks, the 'KEEP' block does not use assignment statements but works with indexing of its two variables 'substmod' and 'indmod'. These can be thought of as vectors that hold the parameters of the substitution and the insertion-deletion model respectively. The user selects the parameters of the substitution model which should be kept fixed by writing 'substmod[index]', where 'index' can be a single non-negative integer that corresponds to the number of the respective parameter or any unordered series of such values with commas as delimiters. Following the conventions of C++, the numbering of the parameters starts with 0. In the case that none of the substitution model parameters should be altered by the program, the user may write 'substmod[all]'. The variable 'indmod' is accessed accordingly.

The 'COOL' block:

This block controls some parameters of the simulated annealing's cooling schedule as well as some options which influence the programs runtime and the number of recorded intermediate alignments. As described in chapter 5, the cooling schedule of the simulated annealing is influenced by two parameters which are estimated from the data, namely the maximum jump of the optimized function in one step (U) and the loglikelihood per position ($\ln L/m$), and by two parameters whose values may be adjusted by the user and which were called c and a in chapter 5. Since both the parameter optimization and the tree optimization have their own cooling schedule, the 'COOL' block has a pair of variables for each of them. For parameter optimization, they are called 'cparam' and 'alphaparam' and for tree optimization, their names are 'ctree' and 'alphatree'. The default value for each of these variables is 0.5. The number of iterations at the beginning of the simulated annealing procedure which are used to estimate U and $\ln L/m$ is held by the variable 'burnin'. The user can also specify the maximum number of iterations (variable 'maxsteps'), the number of iterations after which the program produces output to the screen and to a logfile (variable 'reportafter') and the size of the windows in which pairwise alignments are optimized (variable 'resamprange').

The 'FILES' block:

The program requires two input files and produces four output files. As input, ALIFRITZ expects a file which contains a guide tree in Newick format (Felsenstein, 1993) whose name can be set by an assignment to the variable 'guidetree' and a file with the sequence data in either FASTA (Pearson and Lipman, 1988) or PHYLIP format (Felsenstein, 1993) whose name is held by the variable 'infile'. In its current version, the program does not automatically recognize these two sequence file formats. Therefore, the variable 'format' has to be set to 0 if the sequences are in FASTA format and to 1 if they are PHYLIP formatted. The default format is FASTA. As output, the program creates a file which records intermediate steps of the optimization (variable 'logfile') and three files which hold the tree topology (variable 'outtree'), the parameter values (variable 'outparam') and the multiple alignment (variable 'outalign') of the optimum. The default file names coincide with the names of the variables of the 'FILES' block.

A.2   SIMULATOR

The program SIMULATOR is a software tool which generates data sets of related sequences for a given phylogeny and a given evolutionary model. Unlike currently available simulation programs (Rambaut and Grassly, 1997; Schöniger and von Haeseler, 1995b; Stoye et al., 1998), SIMULATOR does not only offer a wide variety of nucleotide or amino acid substitution models but also a well-defined insertion-deletion dynamics[1]. The program is run from

---

[1]To our knowledge, the program ROSE (Stoye et al., 1998) is the only other program which creates indels. Yet, the process with which these are produced is not strictly model based.

the command line and has to be called with the following three parameters:

1. The name of the control file which holds the settings of the various program options (see below).

2. The expected sequence length.

3. The number of data sets to be generated.

Thus, if we invoke the program with 'simulator ctrl 10000 100', we will get 100 data sets which have evolved according to the settings in the file 'ctrl' and where the expected sequence length is 10000. The control files have the syntax that was described in the last section. The variables are grouped in the three blocks 'SUBST', 'INDEL' and 'FILES'. The block 'SUBST' holds the settings for the substitution model. The models that can be chosen are the very same as the ones which are used by ALIFRITZ. See section A.1 for a detailed description. The variables which describe the desired insertion-deletion process are grouped under the heading 'INDEL'. Again, the choices that can be made coincide with ALIFRITZ's options. The names of the input and output files of the program are specified in the block 'FILES'. The name of the input file that holds the Newick-formatted guide tree is stored in the variable 'treefile'. Two output files are produced by the program. The first one whose name is set through the variable 'alifile' contains the simulated data sets in PHYLIP format. The second one which is represented by the variable 'seqfile' is in FASTA format and lists the simulated sequences in unaligned form. The default file names coincide with the names of the variables in the control file. The program's C++ source code is available on request.

Appendix B

Two useful algorithms

B.1 Sampling pairwise alignments in linear space

The algorithm described in chapter 5 involves steps in which the most probable alignment of two sequences has to be found or in which alignments of two sequences are sampled according to their probability. If we tried to do this via allocating enough memory for the entire alignment graph, we would soon run into trouble, as the alignment graph grows with the product of the sequence lengths. Moreover, even if we restricted the use of our program to sequences of moderate length, its space requirement could still be too high, as we are also performing pairwise alignments of sequences at a phylogeny's interior nodes whose lengths are not known in advance. In the following, we give a short description of an algorithm that allows the sampling of alignments in linear space while only doubling the computing time. It combines Hirschberg's linear-space optimal alignment algorithm (Hirschberg, 1977) and the forward and backward algorithms for pair-HMMs (Durbin *et al.*, 1998). Before describing the algorithm, we will introduce three ideas which are indispensable for its implementation.

The first thing we have to notice is the fact that allocating the whole alignment table is unnecessary if we are only interested in the joint probability of two sequences but not in their alignment. In chapter 4, it has been shown that the values of $f(i, j, \cdot)$ can be computed easily if the values for $f(i-1, j-1, \cdot)$, $f(i-1, j, \cdot)$ and $f(i, j-1, \cdot)$ are known (here the '$\cdot$' stands for any of states $\binom{B}{B}$, $\binom{B}{-}$ or $\binom{-}{B}$). Thus, to compute the entries in the ith row of the alignment table we only need the entries in row $i - 1$. Therefore, the space which we need to compute two sequences' joint probability only grows linearly with the length of the second sequence.

The second building block of our algorithm is the idea that the sequences' joint probability can be computed not only with the forward algorithm from chapter 4 but also in the opposite direction with the following backward algorithm: Given two sequences x and y of lengths n and m, let $b(i, j, v)$ denote the probability that a TKF2 pair-HMM produces a sequence 1 whose last $n - i$ bases coincide with the last $n - i$ bases of x and a second sequence whose last $m - j$ bases coincide with the last $m - j$ bases of y conditioned on the alignment of the two sequences going through the edge (i, j, v) where $v \in \{\binom{B}{B}, \binom{B}{-}, \binom{-}{B}\}$. We initialize the recursion by setting $b(n, m, v) = P(v \rightarrow End)$ and $b(n + 1, j, v) = b(i, m + 1, v) = 0$ for all $i \in \{0, ..., n\}$ and $j \in \{0, ..., m\}$. Then we can perform the following recursion for $i = n, ..., 0$ and $j = m, ..., 0$ except (n, m) and $(0, 0)$:

$$
\begin{aligned}
b(i, j, v) = \ & b(i + 1, j + 1, \tbinom{B}{B}) \cdot P(v \rightarrow \tbinom{B}{B}) \cdot \pi_{x_{i+1}} \cdot P(x_{i+1} \rightarrow y_{j+1}) \\
& + b(i + 1, j, \tbinom{B}{-}) \cdot P(v \rightarrow \tbinom{B}{-}) \cdot \pi_{x_{i+1}} \\
& + b(i, j + 1, \tbinom{-}{B}) \cdot P(v \rightarrow \tbinom{-}{B}) \cdot \pi_{y_{j+1}}
\end{aligned}
$$

The probability to observe the sequences x and y is then simply

$$
\begin{aligned}
p(x,y) \;=\; & b\!\left(1,1,\tbinom{B}{B}\right)\cdot P\!\left(\texttt{Start}\to \tbinom{B}{B}\right)\cdot \pi_{x_1}\cdot P(x_1\to y_1)\\[4pt]
& +\,b\!\left(1,0,\tbinom{B}{\_}\right)\cdot P\!\left(\texttt{Start}\to \tbinom{B}{\_}\right)\cdot \pi_{x_1}\\[4pt]
& +\,b\!\left(0,1,\tbinom{\_}{B}\right)\cdot P\!\left(\texttt{Start}\to \tbinom{\_}{B}\right)\cdot \pi_{y_1}
\end{aligned}
$$

The function b can be used to sample alignments just like we used f in chapter 4 and just like in the case of the forward algorithm, we do not have to allocate memory for the whole alignment graph to compute $p(x,y)$ with this backward algorithm as we can compute the $b(i,\cdot,\cdot)$ from the $b(i+1,\cdot,\cdot)$.

Thirdly, we have to notice that the forward and the backward algorithm can be used not only to sample the entire alignment path, but also to sample subpaths when other parts of the path are already given: Consider the case that we want to sample the alignment path between the nodes (l,r) and (l',r'), where $l<l'$ and $r<r'$, i. e. we want to align the bases $l+1$ to l' of sequence x to the bases $r+1$ to r' of sequence y. Let $a\in\{\texttt{Start},\tbinom{B}{B},\tbinom{B}{\_},\tbinom{\_}{B}\}$ be the last state of the alignment immediately before node (l,r) and let $z\in\{\tbinom{B}{B},\tbinom{B}{\_},\tbinom{\_}{B},\texttt{End}\}$ be the state which follows node (l',r'). In this situation, computing $f(i,j,v)$ or $b(i,j,v)$ for every edge (i,j,v) of the alignment graph becomes redundant. Instead, we compute for every edge in the rectangle defined by the nodes (l,r) and (l',r') the probability that a TKF2 pair-HMM produces a sequence 1 which contains a substring whose first $i-l$ bases coincide with the bases $l+1$ to i of x and a sequence 2 which contains a substring whose first $j-r$ bases coincide with the bases $r+1$ to j of y given that the alignment of the emitted sequences goes through the edge (s,t,a) where s and t are the positions immediately before the respective substrings. We denote this probability with $f_{lrl'r'}(i,j,v\,|\,a)$. It is computed just like $f(i,j,v)$ with the difference that the computation is initialized in the following way:

$$
f_{lrl'r'}\!\left(l+1,r+1,\tbinom{B}{B}\,\big|\,a\right)\;=\;P\!\left(a\to \tbinom{B}{B}\right)\cdot \pi_{x_{l+1}}\cdot P(x_{l+1}\to y_{r+1}),
$$

103

$$f_{lrl'r'}(l+1, r, \tbinom{B}{\_} \mid a) = P(a \to \tbinom{B}{\_}) \cdot \pi_{x_{l+1}},$$

$$f_{lrl'r'}(l, r+1, \tbinom{\_}{B} \mid a) = P(a \to \tbinom{\_}{B}) \cdot \pi_{y_{r+1}} \text{ and}$$

all $f_{lrl'r'}(i, r, \tbinom{B}{B} \mid a) = f_{lrl'r'}(i, r, \tbinom{\_}{B} \mid a) = f_{lrl'r'}(l, j, \tbinom{B}{B} \mid a) = f_{lrl'r'}(l, j, \tbinom{B}{\_} \mid a) = 0$. Likewise, let $b_{lrl'r'}((i, j, v) \wedge z)$ signify the probability that the following events occur in a run of a TKF2 pair-HMM:

- The emitted sequence 1 contains a substring whose last $l' - i$ bases coincide with the bases $i + 1$ to l' of x,

- the emitted sequence 2 contains a substring whose last $r' - j$ bases coincide with the bases $j + 1$ to r' of y,

- the alignment of the emitted sequences goes through the node (s', t'), which is followed by alignment state z, where s' and t' are the last positions of the respective substrings,

conditioned on the alignment of the emitted sequences going through edge $(s' - l' + i, t' - r' + j, v)$, i.e. through the edge which corresponds to the edge (i, j, v) in the alignment of x and y. We can use the same recursion that we used to calculate $b(i, j, v)$ to compute $b_{lrl'r'}((i, j, v) \wedge z)$, but it is now initialized by setting $b_{lrl'r'}((l', r', v) \wedge z) = P(v \to z)$ and all $b(l' + 1, j, v) = b(i, r' + 1, v) = 0$ for all $i \in \{l, ..., l'\}$, $j \in \{r, ..., r'\}$ and $v \in \{\tbinom{B}{B}, \tbinom{B}{\_}, \tbinom{\_}{B}\}$.

The problem of sampling an alignment in linear space can now be solved with the procedure $SAMPAL(l, r, l', r', a, z)$:

- Let $i^* = \frac{l+l'}{2}$ and compute the values of $f_{lrl'r'}(i^*, \cdot, \cdot \mid a)$ and $b_{lrl'r'}((i^*, \cdot, \cdot) \wedge z)$ in linear space as described above.

- For every $j \in \{r, ..., r'\}$ and every $v \in \{\tbinom{B}{B}, \tbinom{B}{\_}, \tbinom{\_}{B}\}$ compute
$p_{lrl'r'}((i^*, j, v) \wedge z \mid a) := f_{lrl'r'}(i^*, j, v \mid a) \cdot b_{lrl'r'}((i^*, j, v) \wedge z)$. Obviously, $p_{lrl'r'}((i^*, j, v) \wedge z \mid a)$ is the probability that a TKF2 pair-HMM

104

produces two sequences where the bases of the first one from some position $s+1$ to some other position s' coincide with the bases $l+1$ to l' of x and where the bases of the second one from some position $t+1$ to some other position t' coincide with the bases $r+1$ to r' of y and whose alignment goes through the edge $(s+i^*-l, t+j-r, v)$ and through the node (s', t') which is followed by state z given that their alignment goes through the edge (s, t, a).

- Sample an edge (i^*, j^*, v^*) according to $\frac{p_{lrl'r'}((i^*,j^*,v^*) \wedge z \,|\, a)}{\sum_{j,v} p_{lrl'r'}((i^*,j,v) \wedge z \,|\, a)}$, i. e. from the posterior distribution of the edges with first component i^* conditioned on the states a and z at the beginning and the end of the subgraph for which we performed the computation.

- Sample edges backwards like in chapter 4 until the sampled edge's first component is $i^* - 1$ and its third component is either $\begin{smallmatrix}\text{B}\\\text{B}\end{smallmatrix}$ or $\begin{smallmatrix}\text{B}\\\text{\_}\end{smallmatrix}$.

- In a similar way, sample forwards until the sampled edge's first component is $i^* + 1$.

- Call $SAMPAL(l, r, i^*-2, j_1, a, z')$ and $SAMPAL(i^*+1, j_2, l', r', a', z)$, where z' is the alignment state of the last edge which was sampled backwards, j_1 is either this edge's second component if $z' = \begin{smallmatrix}\text{B}\\\text{\_}\end{smallmatrix}$ or this edge's second component minus 1 if $z' = \begin{smallmatrix}\text{B}\\\text{B}\end{smallmatrix}$, whereas a' is the alignment state of the last edge that was sampled forwards and j_2 is this edge's second component. See also figure B.1 for an illustration of this recursion.

We begin the computation by calling $SAMPAL(0, 0, n, m, \text{Start}, \text{End})$. The time analysis for this procedure is analogous to the one given on page 258 of Gusfield (1997) for Hirschberg's optimal alignment algorithm. There it is shown that this algorithm only takes twice as long as algorithms that allocate memory for the entire alignment graph.

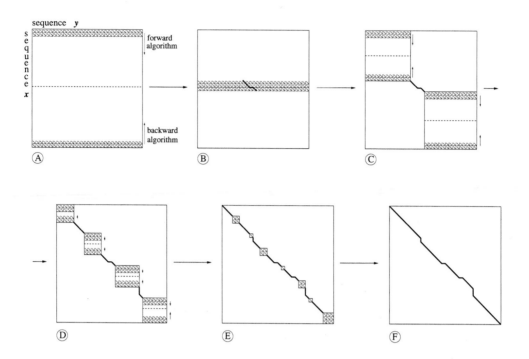

Figure B.1: Sampling alignments in linear space. Ⓐ: In order to compute the $f(\cdot,\cdot,\cdot)$ and the $b(\cdot,\cdot,\cdot)$ values (see text) one does not have to allocate memory for the whole alignment graph whose size is indicated by the rectangle, but only for four of its rows (part of the rectangle where the alignment graph is drawn). The rows which hold the $f(\cdot,\cdot,\cdot)$ values are slid downwards and the ones which hold the $b(\cdot,\cdot,\cdot)$ values move upwards until they meet in the middle of the alignment graph (dashed line). Ⓑ: The stored values are used to sample a subpath of the alignment (thick line). Ⓒ: The sampling of the alignment path then decomposes into sampling a path between the upper left corner of the alignment graph and the beginning of the subpath which has already been sampled, and into sampling a path which goes from this subpath's end to the lower right corner. Ⓓ to Ⓕ: The recursion continues until the whole path is sampled.

B.2 Neighbor joining under constraints

As described in appendix A, the program ALIFRITZ may construct a neighbor-joining tree (Saitou and Nei, 1987) for a set of unaligned sequences. Yet, the user may also provide a tree file which contains the branches that should not be changed by the program. In the following, we will give an outline of a simple method which constructs a neighbor-joining tree under the inclusion of predefined branches.

Let T be an unrooted tree and $D = (d_{ij})$ a distance matrix that holds the pairwise distances of the sequences at the leaves. The tree itself might either be a completely unresolved star, a completely resolved binary tree or anything in between. Our task is to find the binary tree that fits the distance matrix and that includes each of T's branches. Let every leaf node of T be labelled with the index of the row of D that holds its distances to the other leaves and let all the internal nodes be unlabelled. We start by picking randomly one of the leaves as root of the tree and then call for this node the procedure $NEIGHBOUR$:

- For every child, i. e. for every adjacent node that does not lie towards the root, call the procedure $NEIGHBOUR$.

- If the calling node, i. e. the node from which the procedure $NEIGHBOUR$ was invoked, has at least two children, do:

  - compute for every child node i its net divergence (r_i) from the other labelled nodes: $r_i = \sum_{k=1}^{N} d_{ik}$, where N is the number of columns of the distance matrix;

  - for every pair of child nodes i and j, calculate a rate corrected distance $m_{ij} := d_{ij} - (r_i + r_j)/(M - 2)$; the variable M was set equal to the number of taxa at the start of the recursion;

107

- find the pair of children i^* and j^* whose corrected distance is minimal;

- if the calling node has more than two children, then introduce a new node which is the parent of i^* and j^* but a child of the calling node;

- set v_{pi^*} — the length of the branch between i^* and its parent p — equal to $d_{i^*j^*}/2 + (r_i^* - r_j^*)/[2(M-2)]$ and set v_{pj^*} — the length of the branch between j^* and the parent node — equal to $d_{i^*j^*} - v_{pi^*}$;

- label the parent node with the number of columns of D plus 1;

- add the distance between any labelled node k and p to every row or column k of D; this distance is defined as follows: $d_{kp} := (d_{i^*k} + d_{j^*k} - d_{i^*j^*})/2$; set every entry of D where either the row or the column number is equal to i^* or to j^* as well as d_{pp} equal to 0;

- decrement M by 1;

while the calling node has more than two children.

- If the calling node is the root, then set the length of the branch between the root and its child equal to d_{rs}, where r is the label of the root node and s the one of its child.

The combination of the if-condition and of the termination condition of the do-$while$-loop in the above procedure might look peculiar. However, both of them are necessary: We want the do-$while$-loop to be executed until the subtree at the calling node is completely resolved and until we know the lengths of the branches between the calling node and its two remaining children. Hence the loop's termination condition. On the other hand, we do not want

108

to enter the *do-while*-loop if the calling node has less than two children, as its computations are not defined in this situation. Hence the if-condition.

Appendix C

Generalizations of the TKF2 model

When applying the TKF2 model (see chapter 5), we make several assumptions about the insertion-deletion process. It should be noted, however, that some of them are expendable. In the next sections, we will show how to compute the joint probability of two sequences when the fragments are not drawn from one geometric distribution but from a mixture of geometric distributions (C.2) and we will explain what to do if the distribution of fragment lengths is arbitrary (C.1).

C.1 Insertion-deletion models with arbitrary distributions of fragment lengths

Think of two sequences x and y of length n and m, respectively. Consider x to be ancestral to y. Like in the TKF2 model, let these sequences be composed of fragments which are never split by insertions and which are removed as a whole if hit by a deletion. Let insertions between any two fragments and at the ends of the sequence occur at rate λ and let μ be the deletion rate of the fragments. Keep also the notation convention of the TKF2 model. The only thing we change is the distribution of fragment lengths: While the probability that a fragment has length l, l being at least 1, is $\rho^{l-1}(1-\rho)$ in the TKF2 model, the distribution can now have any shape. Clearly, the probability of fragment survival (α), the probability of more extant descendants of a fragment given that there is at least one (β) and the probability of a fragment having extant descendants given that it did not survive (γ) can be computed with the same expressions that were used in chapter 4 for the TKF1 model. However, dropping the assumption of geometrically distributed fragment lengths destroys the Markov property of the alignment positions. Nevertheless, the following recursion makes it possible to compute the joint probability of x and y.

Let $\frac{F}{F}$ denote the situation that a fragment of any length has survived. Similarly, let $\frac{F}{\_}$ symbolize the deletion and $\frac{\_}{F}$ the insertion of an arbitrarily long fragment. Let $g(i, j, \frac{F}{F})$ be the probability that a realization of our insertion-deletion process produces two sequences which fulfill the subsequent conditions.

- The first i bases of the ancestral sequence coincide with the first i bases of x.

- The first j bases of the descendant sequence coincide with the first j bases of y.

- There is a fragment boundary immediately after position i of the ancestral sequence.

- The fragment which ends in the ancestral sequence's ith position is homologous to a substring of the descendant sequence which ends with

112

this sequence's jth position. As fragment boundaries are not changed by the insertion-deletion model this implies that this substring is also one entire fragment. We will say that the fragment ends in node (i, j) of the alignment graph.

$g(i, j, \frac{F}{\_})$ and $g(i, j, \frac{\_}{F})$ are defined similarly, except that the fragment which ends in position i of sequence 1 is aligned to gaps between the jth and the $j + 1$th site of sequence 2 and that the fragment that ends with site j of sequence 2 is aligned to gaps between the position i and $i + 1$ of sequence 1, respectively. If a fragment has survived and ends in node (i, j) we enter (i, j) through the edge $(i, j, \frac{B}{B})$. Thus, the fragment can have any length from 1 up to the minimum of i and j. If the fragment got deleted, i. e. we enter (i, j) through the edge $(i, j, \frac{B}{\_})$, the maximum fragment length is equal to i, and if it was inserted it may at most comprise j positions. Therefore, the following equations hold

for $i = 1, ..., n$ and $j = 1, ..., m$

$$g(i, j, \tfrac{F}{F}) = \sum_{k=1}^{min(i,j)} [g(i - k, j - k, \tfrac{F}{F}) \cdot (1 - \beta) + g(i - k, j - k, \tfrac{F}{\_}) \cdot (1 - \gamma)$$

$$+ g(i - k, j - k, \tfrac{\_}{F}) \cdot (1 - \beta)] \cdot \alpha \cdot \frac{\lambda}{\mu} \cdot \pi_{F_k} \cdot \prod_{h=0}^{k-1} \pi_{x_{i-h}} Pr(x_{i-h} \to y_{j-h}),$$

for $i = 1, ..., n$ and $j = 0, ..., m$

$$g(i, j, \tfrac{F}{\_}) = \sum_{k=1}^{i} [g(i - k, j, \tfrac{F}{F}) \cdot (1 - \beta) + g(i - k, j, \tfrac{F}{\_}) \cdot (1 - \gamma) + g(i - k, j, \tfrac{\_}{F})$$

$$\cdot (1 - \beta)] \cdot (1 - \alpha) \cdot \frac{\lambda}{\mu} \cdot \pi_{F_k} \cdot \prod_{h=0}^{k-1} \pi_{x_{i-h}}$$

and for $i = 0, ..., n$ and $j = 1, ..., m$

$$g(i, j, \tfrac{\_}{F}) = \sum_{k=1}^{j} [g(i, j - k, \tfrac{F}{F}) \cdot \beta + g(i, j - k, \tfrac{F}{\_}) \cdot \gamma + g(i, j - k, \tfrac{\_}{F}) \cdot \beta] \cdot \pi_{F_k}$$

$$\cdot \prod_{h=0}^{k-1} \pi_{y_{j-h}},$$

113

where π_{F_k} is the probability of a fragment of size k. We initialize the recursion by setting $g(0,0,{}^F_F) = 1$, $g(0,0,{}^F_-) = 0$ and $g(0,0,{}_F^-) = 0$. The probability to observe the sequences x and y is then given by

$$p(x,y) = \left[g(n,m,{}^F_F) \cdot (1-\beta) + g(n,m,{}_F^-) \cdot (1-\beta) + g(n,m,{}^F_-) \cdot (1-\gamma) \right] \\ \cdot (1 - \lambda/\mu) .$$

The expectation-maximization algorithm described by Thorne and Churchill (1995) can be as easily adapted to this recursion as can the alignment sampling procedure from chapter 4. Thus, we can now get maximum likelihood estimates of the model parameters and sample alignment paths from their posterior distribution without strong assumptions about the size distribution of indels. However, like in the case of optimal alignment with arbitrary gap weights (cf. Gusfield, 1997, pp. 241), the computation time for our algorithm grows with $O(nm^2 + mn^2)$. Thus, it will be important to identify those distributions which permit faster recursions. The next section will describe one of these. Other obvious examples are the cases in which fragments have constant size or — more reasonably — where the fragment length may not exceed a certain number of positions.

114

C.2 Insertion-deletion models with several classes of fragments

Here, just like in the last section, we will keep all the features of the TKF2 model except for the geometric distribution of fragment lengths. Therefore, α is still the probability of fragment survival, β the probability of more extant descendants of a fragment given that there is at least one, and γ the probability of a fragment having extant descendants given that it did not survive, All these probabilities can still be computed from the insertion rate λ and the deletion rate μ like in chapter 4. We just assume now that the lengths of the sequence fragments are drawn from a mixture of k geometric distributions, i. e. the probability that a fragment has some length $l \geq 1$ is equal to $\sum_{h=1}^{k} w_h \rho_h^{l-1}(1-\rho_h)$, where the ρ_h are probabilities and the weights w_h of the individual geometric distributions are positive and sum up to 1. Thus, every fragment chooses with probability w_h the distribution $\rho_h^{l-1}(1-\rho_h)$ from which its length should be drawn. Although this scenario might seem artificial, it is quite a natural setting if we consider the manifold causes for insertions or deletions. Using this approach we can for example have a class of fragments that has a small expected length and thus mimics insertions and deletions via slippage, and we may have a class of fragments with a bigger expected length that stands for the insertions or deletions of longer stretches of DNA. As every fragment has to pick its geometric distribution, the weights w_h and the various parameters ρ_h of these distributions have to be taken care of when computing the transition probabilities for this model. Therefore, a pair-HMM with the states Start, $\begin{smallmatrix} B \\ B \end{smallmatrix}$, $\begin{smallmatrix} B \\ \_ \end{smallmatrix}$, $\begin{smallmatrix} \_ \\ B \end{smallmatrix}$ and End is not sufficient. Instead, each of the k geometric distributions that are involved in the mixed distribution now has to have its own state for homologous positions, insertions and deletions. Let

${}^{B}_{Bh}$, $\bar{{}_{Bh}}$ and ${}^{B}_{-h}$ symbolize the respective states for the hth geometric distribution. With this enlarged state space and under the assumption that the ancestral sequence is drawn from the equilibrium distribution of sequences, our model can be represented as a pair-HMM whose transition probabilities are shown in table C.1. The forward and backward algorithms from chapter 4 and appendix B can easily be modified so that they can be applied to this pair-HMM. Thus, the computation of the joint probability of two sequences has the same time and space bounds under this scenario as under the simpler TKF2 model. Yet, the time it takes to compute each entry of the alignment table grows quadratically with the number of fragment classes. This number will, however, be rather small in any reasonable application of this model.

	$y = \begin{smallmatrix}\mathrm{B}\\\mathrm{B}h\end{smallmatrix}$	$y = \begin{smallmatrix}\mathrm{B}\\-h\end{smallmatrix}$	$y = \begin{smallmatrix}\\\bar{\mathrm{B}}h\end{smallmatrix}$	$y = \mathbf{End}$
$x = \mathbf{Start}$	$(1-\beta)\frac{\lambda}{\mu}\alpha w_h$	$(1-\beta)\frac{\lambda}{\mu}(1-\alpha)w_h$	βw_h	$(1-\beta)(1-\frac{\lambda}{\mu})$
$x = \begin{smallmatrix}\mathrm{B}\\\mathrm{B}g\end{smallmatrix}$	if $g = h$: $$\frac{\rho_g + (1-\rho_g)(1-\beta)\frac{\lambda}{\mu}\alpha w_g}{}$$ if $g \neq h$: $(1-\rho_g)(1-\beta)\frac{\lambda}{\mu}\alpha w_h$	$(1-\rho_g)(1-\beta)\frac{\lambda}{\mu}(1-\alpha)w_h$	$(1-\rho_g)\beta w_h$	$(1-\rho_g)(1-\beta)(1-\frac{\lambda}{\mu})$
$x = \begin{smallmatrix}\mathrm{B}\\-g\end{smallmatrix}$	$(1-\rho_g)(1-\gamma)\frac{\lambda}{\mu}\alpha w_h$	if $g = h$: $$\frac{\rho_g + (1-\rho_g)(1-\gamma)\frac{\lambda}{\mu}(1-\alpha)w_g}{}$$ if $g \neq h$: $(1-\rho_g)(1-\gamma)\frac{\lambda}{\mu}(1-\alpha)w_h$	$(1-\rho_g)\gamma w_h$	$(1-\rho_g)(1-\gamma)(1-\frac{\lambda}{\mu})$
$x = \begin{smallmatrix}\\\bar{\mathrm{B}}g\end{smallmatrix}$	$(1-\rho_g)(1-\beta)\frac{\lambda}{\mu}\alpha w_h$	$(1-\rho_g)(1-\beta)\frac{\lambda}{\mu}(1-\alpha)w_h$	if $g = h$: $$\frac{\rho_g + (1-\rho_g)\beta w_g}{}$$ if $g \neq h$: $(1-\rho_g)\beta w_h$	$(1-\rho_g)(1-\beta)(1-\frac{\lambda}{\mu})$

Table C.1: The transition probabilities $P(x \rightarrow y)$ of the pair-HMM for the fragment insertion-deletion model when the fragments are drawn from a mixture of geometric distributions. See chapter 4 on how to compute α, β and γ. The indices g and h may take any value between 1 and the number of geometric distributions that constitute the mixed distribution.

Bibliography

Adachi, J. and Hasegawa, M. 1996. Model of amino acid substitutions in proteins encoded by mitochondrial DNA. *J. Mol. Evol.* **42**:459–468.

Agarwal, P. and States, D. J. 1996. A Bayesian evolutionary distance for parametrically aligned sequences. *J. Comp. Biol.* **3**:1–17.

Altschul, S. and Lipman, D. J. 1989. Trees, stars, and multiple biological sequence alignment. *SIAM Journal of applied mathematics* **49**:197–209.

Altschul, S. F. and Erickson, B. W. 1986. Optimal sequence alignment using affine gap costs. *Bull. Math. Biol.* **48**:603–616.

Barton, G. J. and Sternberg, M. J. E. 1987. A strategy for the rapid multiple alignment of protein sequences. *J. Mol. Biol.* **198**:327–337.

Bellman, R. E. 1957. *Dynamic Programming.* Princeton: Princeton University Press.

Berger, M. P. and Munson, P. J. 1991. A novel randomized iterative strategy for aligning multiple protein sequences. *CABIOS* **7**:479–484.

Bishop, M. and Thompson, E. 1986. Maximum likelihood-alignment of DNA sequences. *J. Mol. Biol.* **190**:159–165.

Briffeuil, P., Baudoux, G., Lambert, C., Boile, X. D., Vinals, C., Feytmans, E., and Depiereux, E. 1998. Comparative analysis of seven multiple protein sequence alignment servers: clues to enhance reliability of predictions. *Bioinformatics* **14(4)**:357–366.

Bunemann, P. 1971. The recovery of trees from measures of dissimilarity. In: *Mathematics in the archeological and historical sciences* (F. R. Hodson, D. G. Kendall, P. T., ed.) pp. 387–395. Edinburgh University Press.

Burckhardt, F., von Haeseler, A., and Meyer, S. 1999. HvrBase: Compilation of mtDNA control region sequences from primates. *Nucl. Acids Res.* **27**:138–142.

Carrillo, H. and Lipman, D. J. 1988. The multiple sequence alignment problem in biology. *SIAM Journal of applied mathematics* **48**:1073–1082.

Cavalli-Sforza, L. L. and Edwards, A. W. F. 1967. Phylogenetic analysis: Models and estimation procedures. *Am. J. Hum. Genet.* **19**:233–257.

Dayhoff, M. O., ed. 1978. *Atlas of Protein Sequences and Structure*, volume 5. Silver Springs: Natl. Biomed. Res. Found.

de Finetti, B. 1926. Considerazioni matematiche sull'ereditarietà mendeliana. *Metron* **6**:29–37.

Durbin, R., Eddy, S., Krogh, A., and Mitchison, G. 1998. *Biological sequence analysis: Probabilistic models of proteins and nucleic acids.* Cambridge: Cambridge University Press.

Edwards, A. W. F. 1972. *Likelihood.* Cambridge: Cambridge University Press.

Ewens, W. J. and Grant, G. R. 2001. *Statistical methods in bioinformatics.* New York: Springer-Verlag.

Farris, J. S., Kluge, A. G., and Eckhardt, M. J. 1970. A numerical approach to phylogenetic systematics. *Syst. Zool.* **19**:172–191.

Felsenstein, J. 1978a. The number of evolutionary trees. *Syst. Zool.* **27**:27–33.

Felsenstein, J. 1978b. Cases in which parsimony or compatibility methods will be positively misleading. *Syst. Zool.* **27**:401–410.

Felsenstein, J. 1981. Evolutionary trees from DNA sequences: A maximum likelihood approach. *J. Mol. Evol.* **17**:368–376.

Felsenstein, J. 1993. *PHYLIP: Phylogenetic Inference Package, Version. 3.5c.* Seattle: Department of Genetics, University of Washington.

Feng, D.-F. and Doolittle, R. F. 1987. Progressive sequence alignment as a prerequisite to correct phylogenetic trees. *J. Mol. Evol.* **25**:351–360.

Fitch, W. M. 1971. Toward defining the course of evolution: Minimum change for a specific tree topology. *Syst. Zool.* **20**:406–416.

Fitch, W. M. and Margoliash, E. 1967. Construction of phylogenetic trees. *Science* **155**:279–284.

Fitch, W. M. and Smith, T. F. 1983. Optimal sequence alignments. *Proc. Natl. Acad. Sci. USA* **80**:1382–1386.

Gamerman, D. 1997. *Markov Chain Monte Carlo.* London: Chapman & Hall.

Gonnet, G. H., Korostensky, C., and Benner, S. 2000. Evaluation measures of multiple sequence alignments. *J. Comp. Biol.* **7(1/2)**:261–276.

Gotoh, O. 1982. An improved algorithm for matching biological sequences. *J. Mol. Biol.* **162**:705–708.

Gotoh, O. 1996. Significant improvement in accuracy of multiple protein alignments by iterative refinement as assessed by reference to structural alignments. *J. Mol. Biol.* **264**:823–838.

Gotoh, O. 1999. Multiple sequence alignments: algorithms and applications. *Adv. Biophys.* **36**:159–206.

Gribskov, M., McLachlan, A., and Eisenberg, D. 1987. Profile analysis detection of distantly related proteins. *Proc. Natl. Acad. Sci. USA* **88**:4355–4358.

Gu, X. and Li, W.-H. 1995. The size distribution of insertions and deletions in human and rodent pseudogenes suggest the logarithmic gap penalty for sequence alignment. *J. Mol. Evol.* **40**:464–473.

Gusfield, D. 1997. *Algorithms on Strings, Trees, and Sequences: Computer Science and Computational Biology*. Cambridge: Cambridge University Press.

Handt, O., Meyer, S., and von Haeseler, A. 1998. Compilation of human mtDNA control region sequences. *Nuc. Acids Res.* **26**:126–130.

Hasegawa, M., Kishino, H., and Yano, K. 1985. Dating of the human-ape splitting by a molecular clock of mitochondrial DNA. *J. Mol. Evol.* **22**:160–174.

Hein, J., Wiuf, C., Knudsen, B., Møller, M. B., and Wibling, G. 2000. Statistical alignment:computational properties, homology testing and goodness-of-fit. *J. Mol. Biol.* **302**:265–279.

Henikoff, S. and Henikoff, J. G. 1992. Amino acid substitution matrices from protein blocks. *Proc. Natl. Acad. Sci. USA* **89**:10915–10919.

Higgins, D. G. and Sharp, P. M. 1989. Fast and sensitive multiple sequence alignments on a microcomputer. *CABIOS* **5**:151–153.

Hirschberg, D. S. 1977. Algorithms for the longest common subsequence problem. **24**:664–675.

Holmes, I. and Bruno, W. J. 2001. Evolutionary HMMs: a Bayesian approach to multiple alignment. *Bioinformatics* **17**:803–820.

Holmes, I. and Durbin, R. 1998. Dynamic programming alignment accuracy. *J. Comp. Biol.* **5**:493–504.

Huelsenbeck, J. P. 1995. Performance of phylogenetic methods in simulation. *Syst. Biol.* **44**:17–48.

Huelsenbeck, J. P. and Hillis, D. M. 1993. Success of phylogenetic methods in the four-taxon case. *Syst. Biol.* **42**:247–264.

Hwa, T. and Lässig, M. 1996. Similarity detection and localization. *Phys. Rev. Lett.* **76**:2591–2594.

Jones, D. T., Taylor, W. R., and Thornton, J. M. 1992. The rapid generation of mutation data matrices from protein sequences. *CABIOS* **8**:275–282.

Jukes, T. H. and Cantor, C. R. 1969. Evolution of protein molecules. In: *Mammalian Protein Metabolism* (Munro, H. N., ed.) pp. 21–132. Academic Press New York.

Kimura, M. 1980. A simple method for estimating evolutionary rates of base substitutions through comparative studies of nucleotide sequences. *J. Mol. Evol.* **16**:111–120.

Kimura, M. 1981. Estimation of evolutionary distances between homologous nucleotide sequences. *Proc. Natl. Acad. Sci. USA* **78**:454–458.

Lake, J. A. 1991. The order of sequence alignment can bias the selection of tree topology. *Mol. Biol. Evol.* **8**:378–385.

Lanave, C., Preparata, G., Saccone, C., and Serio, G. 1984. A new method for calculating evolutionary substitution rates. *J. Mol. Evol.* **20**:86–93.

Lipman, D. J., Altschul, S. F., and Kececioglu, J. D. 1989. A tool for multiple sequence alignment. *Proc. Natl. Acad. Sci. USA* **86**:4412–4415.

Lunter, G., Miklós, I., Song, Y. S., and Hein, J. 2003. An efficient algorithm for statistical multiple alignment on arbitrary phylogenetic trees. *J. Comp. Biol.* in press.

Mau, B. and Newton, M. A. 1997. Phylogenetic inference for binary data on dendograms using markov chain monte carlo. *J. Comp. Graph. Stat.* **6**:122–131.

McClure, M., Vasi, T., and Fitch, W. 1994. Comparative analysis of multiple protein-sequence alignment methods. *Mol. Biol. Evol.* **11**:571–592.

McGuire, G., Denham, M. C., and Balding, D. J. 2001. Models of sequence evolution for DNA sequences containing gaps. *Mol. Biol. Evol.* **18**:481–490.

Metzler, D. 2003. Statistical alignment based on fragment insertion and deletion models. *Bioinformatics* **19**:490–499.

Mevissen, H. T. and Vingron, M. 1996. Quantifying the local reliability of a sequence alignment. *Protein Engineering* **9**:127–132.

Miklós, I. and Toroczkai, Z. 2001. An improved model for statistical alignment. In: *Algorithms in bioinformatics* (Gascuel, O. and Moret, B. M. E., eds.) pp. 1–10. Springer Berlin, Germany.

Miller, W. and Myers, E. W. 1988. Sequence comparison with concave weighting functions. *Bull. Math. Biol.* **50**:97–120.

Morrison, D. A. and Ellis, J. T. 1997. Effects of Nucleotide Sequence Alignment on Phylogeny Estimation: A case study of 18S rDNAs of Apicomplexa. *Mol. Biol. Evol.* **14**:428–441.

Müller, T. and Vingron, M. 2000. Modeling amino acid replacement. *J. Comp. Biol.* **7**:761–776.

Needleman, S. B. and Wunsch, C. D. 1970. A general method applicable to the search for similarities in the amino acid sequence of two proteins. *J. Mol. Biol.* **48**:443–453.

Olsen, G. J., Natsuda, H., Hagstrom, R., and Overbeek, R. 1994. FastDNAML: A tool for construction of phylogenetic trees of DNA sequences using maximum likelihood. *CABIOS* **10**:41–48.

Pearson, W. R. and Lipman, D. J. 1988. Improved tools for biological sequence comparison. *Proc. Natl. Acad. Sci. USA* **85**:2444–2448.

Penny, D., Foulds, L. R., and Hendy, M. D. 1982. Testing the theory of evolution by comparing phylogenetic trees constructed from five different protein sequences. *Nature* **297**:197–200.

Rambaut, A. and Grassly, N. C. 1997. Seq-Gen: An application for the Monte Carlo simulation of DNA sequence evolution along phylogenetic trees. *CABIOS* **13**:235–238.

Robinson, D. F. and Foulds, L. R. 1981. Comparison of phylogenetic trees. *Math. Biosci.* **53**:131–147.

Rogers, J. S. 1997. On the consistency of maximum likelihood estimation of phylogenetic trees from nucleotide sequences. *Syst. Biol.* **46**:354–357.

Rzhetsky, A. and Nei, M. 1992. A simple method for estimating and testing minimum-evolution trees. *Mol. Biol. Evol.* **9**:945–967.

Saitou, N. and Nei, M. 1987. The neighbor-joining method: A new method for reconstructing phylogenetic trees. *Mol. Biol. Evol.* **4**:406–425.

Saitou, N. and Ueda, S. 1994. Evolutionary rates of insertion and deletion in noncoding nucleotide sequences of primates. *Mol. Biol. Evol.* **11**:504–512.

Salter, L. A. and Pearl, D. K. 2001. Stochastic search strategy for estimation of maximum likelihood phylogenetic trees. *Syst. Biol.* **50**:7–17.

Schöniger, M. and von Haeseler, A. 1993. A simple method to improve the reliability of tree reconstructions. *Mol. Biol. Evol.* **10**:471–483.

Schöniger, M. and von Haeseler, A. 1995a. Performance of the maximum-likelihood, neighbor-joining, and maximum-parsimony methods when sequence sites are not independent. *Syst. Biol.* **44**:533–547.

Schöniger, M. and von Haeseler, A. 1995b. Simulating efficiently the evolution of DNA sequences. *CABIOS* **11**:111–115.

Siegmund, D. and Yakir, B. 2000. Approximate p-values for local sequence alignments. *Ann. Stat.* **28**:657–680.

Smith, T. F. and Waterman, M. S. 1981. Identification of common molecular subsequences. *J. Mol. Biol.* **147**:195–197.

Sonnhammer, E. L. and Durbin, R. 1996. A dot-matrix program with dynamic threshold control suited for genomic dna and protein sequence analysis. *Gene* **167**:1–10.

Steel, M. and Hein, J. 2001. Applying the Thorne-Kishino-Felsenstein model to sequence evolution on a star shaped tree. *Applied Mathematics Letters* **14**:679–684.

Stoye, J., Evers, D., and Meyer, F. 1998. Rose: generating sequence families. *Bioinformatics* **14**:157–163.

Strimmer, K. 1997. *Maximum Likelihood Methods in Molecular Phylogenetics*. München: Herbert Utz Verlag. PhD Thesis.

Strimmer, K. and von Haeseler, A. 1996. Quartet puzzling: A quartet maximum likelihood method for reconstructing tree topologies. *Mol. Biol. Evol.* **13**:964–969.

Swofford, D. L. 1991. *PAUP*: Phylogenetic Analysis Using Parsimony and Other Methods*. Sunderland: Sinauer Associates.

Swofford, D. L., Olsen, G. J., Waddell, P. J., and Hillis, D. M. 1996. Phylogenetic inference. In: *Molecular Systematics* (Hillis, D. M., Moritz, C., and Mable, B. K., eds.) pp. 407–514. Sinauer Associates Sunderland.

Tamura, K. and Nei, M. 1993. Estimation of the number of nucleotide substitutions in the control region of mitochondrial DNA in humans and chimpanzees. *Mol. Biol. Evol.* **10**:512–526.

Tavaré, S. 1986. Some probabilistic and statistical problems in the analysis of DNA sequences. In: *Some Mathematical Questions in Biology: DNA Sequence Analysis* (Waterman, M. S., ed.) pp. 57–86. The American Mathematical Society Providence, Rhode Island.

Thompson, J. D., Higgins, D. G., and Gibson, T. J. 1994. CLUSTAL W: Improving the sensitivity of progressive multiple alignment through sequence weighting, positions-specific gap penalties and weight matric choice. *Nucleic Acids Research* **22**:4673–4680.

Thompson, J. D., Plewniak, F., and Poch, O. 1999. A comprehensive comparison of multiple sequence alignment programs. *Nucleic Acids Research* **27**:2682–2690.

Thorne, J., Kishino, H., and Felsenstein, J. 1991. An evolutionary model for maximum likelihood-alignment of DNA sequences. *J. Mol. Evol.* **33**:114–124.

Thorne, J. L. and Churchill, G. A. 1995. Estimation and reliability of molecular sequence alignments. *Biometrics* **51**:100–113.

Thorne, J. L. and Kishino, H. 1992. Freeing phylogenies from artifacts of alignment. *Mol. Biol. Evol.* **9**:1148–1162.

Thorne, J. L., Kishino, H., and Felsenstein, J. 1992. Inching toward reality: An improved likelihood model of sequence evolution. *J. Mol. Evol.* **34**:3–16.

Voet, D. and Voet, J. G. 1995. *Biochemistry.* New York: Wiley.

Waterman, M. S. 1995. *Introduction to Computational Biology: Maps, Sequences and Genomes.* London: Chapman & Hall.

Waterman, M. S., Eggert, M., and Lander, E. 1992. Parametric sequence comparison. *Proc. Natl. Acad. Sci. USA* **89**:6090–6093.

Waterman, M. S. and Perlwitz, M. D. 1984. Line geometries for sequence comparisons. *Bull. Math. Biol.* **46**:567–577.

Zharkikh, A. 1994. Estimation of evolutionary distances between nucleotide sequences. *J. Mol. Evol.* **39**:315–329.

Zharkikh, A. and Li, W.-H. 1993. Inconsistency of the maximum-parsimony method: The case of five taxa with a molecular clock. *Syst. Biol.* **42**:113–125.